D0064662

SOLID STATE PHYSICS

VOLUME 53

Front Row from left to right: David Turnbull, Frederick Seitz
Back Row from left to right: Henry Ehrenreich, Frans Spaepen

SOLID STATE PHYSICS

Advances in
Research and Applications

Editors

HENRY EHRENREICH

FRANS SPAEPEN

Division of Engineering and Applied Sciences
Harvard University
Cambridge, Massachusetts

VOLUME 53
1955–1999: Overview, Contents and Authors

ACADEMIC PRESS

San Diego   San Francisco   New York   Boston
London   Sydney   Tokyo

ACADEMIC PRESS
525 B Street, Suite 1900, San Diego, CA 92101-4495, USA
http://www.apnet.com

Academic Press
24–28 Oval Road, London NW1 7DX, UK
http://www.hbuk.co.uk/ap/

International Standard Serial Number: 0081-1947
International Standard Book Number: 0-12-607752-5
Printed in the United States of America
99 00 01 02 03 MB 9 8 7 6 5 4 3 2 1

P/O 166034   01-SEP-99

Contents

I. Introduction

I. Introduction

The Solid State Physics series, founded by Frederick Seitz and David Turnbull in 1955, is certainly the oldest review series in the field. It was so widely recognized that, in its early days, it even acquired a nickname, the "Seitzschrift." The series is an outgrowth of Seitz's treatise, *The Modern Theory of Solids*, published in 1940, which played a major role in defining the then fledgling discipline of solid state physics. The field grew sufficiently rapidly during the post-World War II period, that by the time of the mid-1950s, as Seitz notes in the Introduction to Volume 47, marking the retirement of Turnbull as a series editor, "it became clear that the time was again right to provide something in the nature of a comprehensive overview of the field." He goes on to point out that "several individuals suggested that I consider a new edition," but rejects that possibility on the grounds that "any such endeavor would be truly massive if it was to be comprehensive" and furthermore that "the field was in a dynamic state of development and hence was open-ended." He and Turnbull discussed various alternatives during 1954, focusing "on the feasibility of establishing a series of volumes devoted to solid state physics that could serve the needs of the growing group of professional scientists involved in the field and be sufficiently flexible that it could be regarded as open-ended." The series was visualized as "books rather than a review journal in order to provide some element of cohesion," in short, "something intermediate between what are commonly called 'advances' that have a year-to-year characteristic and a single comprehensive text."

The Preface to the first volume stated the intent and goals explicitly. What was said then bears repeating, for it has and will continue to define the contents of these books:

> The viewpoints and activities in certain closely allied fields, particularly electronics, metallurgy, crystallography and chemistry of solids, have been influenced markedly by developments in solid state science. As a result of this expansion of knowledge solid state physicists are finding that, in order to make significant contributions, it is necessary to concentrate their efforts. . . . Because of this specialization it is desirable that a mechanism exist whereby investigators and students can readily obtain a balanced view. . . . The purpose of the present series is to fulfill this need, at least in part, by publication of compact and authoritative reviews of the important areas of the field. . . . Three general types of articles are solicited: (1) broad elementary surveys that have particular value in orienting the advanced graduate student or an investigator having little

previous knowledge of the subject; (2) broad surveys of fields of advanced research that serve to inform and stimulate the more experienced investigators; and (3) more specialized articles describing important new techniques, both experimental and theoretical.

The interdisciplinarity alluded to here, which counterpoises increased research specialization on the one hand, with the need to be ever more broadly informed on the other, has ever increasing influence on the condensed matter and materials science fields. The present editors of the series are fully aware of their responsibility in continuing to attract authoritative and well-written reviews that reflect these broadened scopes of interest.

The present editors, Henry Ehrenreich and Frans Spaepen, have served since 1968 and 1994, respectively. Frederick Seitz relinquished his official responsibilities in 1984 and David Turnbull did the same in 1994.

The present conspectus of the Solid State Physics series, covering the years 1955 to 1999 (Volumes 1–51), is intended to be more than an index. As it covers almost 45 years, it represents an essentially coherent account of some of the principal developments of the field during its time of most rapid growth. It will therefore serve as a convenient reference for those needing to locate specific information or others looking for a didactic introduction to a given area. It can also be viewed as an itemized collection that presents a schematic historical overview of the field.

This book has therefore been organized to meet these different needs.

The Overview

This introductory section provides an overview of some of the developments of key areas in the solid state area together with references to some of the articles in the series as illustrations.

The Subject Index

In keeping with the pedagogical orientation of the series, subjects are arranged much as they would be in an extensive text or treatise covering the field. Each category lists all the pertinent articles in the series. An article appropriate to several headings is listed multiply. However, every review and supplement volume is referenced at least once. Moreover, the list for each category is arranged chronologically in order to provide historical perspective. The primary contributions for a given time period, as cited in the references, are thus also more easily accessible.

The Author Index

This index is given an encyclopedic format. Each article title and the table of contents preceding each in the published text is listed in this index under the name of the first author. Every author is cross-referenced in the case of reviews having multiple authors. A *précis* of each article or book is therefore available to the user, which should help substantially in locating reference or pedagogically oriented material.

We are grateful to Ms. Christine M. Mullaney for her indispensable help in preparing this volume, and to Professor David Turnbull for a critical reading of the manuscript.

HENRY EHRENREICH
FRANS SPAEPEN

II. Overview

II. Overview

Materials Synthesis

The development of fabrication techniques that could produce well-characterized materials has often been the critical step that made reliable solid state experiments possible. Single crystals, with the lowest possible concentrations of lattice defects and impurities, were essential for the development of semiconductor physics and the study of Fermi surfaces. That much of the early fundamental work on the kinetics of phase transformations was performed on metallic alloys, rather than on ceramics or polymers, was no doubt the result of metallurgical synthesis being the most advanced at the time.

The development of zone refining by Pfann (Vol. 4, 1957) was a major advance in the fabrication of chemically pure crystals. As the method was ideally compatible with single crystal growth techniques introduced earlier by Bridgman, Pfann's work was key to the invention of the transistor and became the basis of semiconductor fabrication. Another important electronic material, single crystal quartz, became available on a large scale by the development of the high pressure hydrothermal growth technique, as described by Laudise and Nielsen (Vol.12, 1961). The energetics and kinetics of the atomistic mechanisms that govern the crystal growth process have been reviewed by Parker (Vol. 25, 1970).

More specialized topics include the use of metal catalysts in the high-pressure synthesis of diamond (Bundy and Strong, Vol. 13, 1962); the synthesis of liquid crystals (Keller and Liebert, Suppl. 14, 1978); the formation of small metal clusters from the vapor using nozzles or gas aggregation (de Heer *et al.*, Vol. 40, 1987); and the preparation of fullerenes, which was revolutionized by the Krätschmer-Huffman graphite arc-vaporization technique that enabled the low-cost production of large quantities of these materials (Lieber and Chen, Vol. 48, 1994).

Experimental Techniques

Von Laue's discovery in 1912 that crystals diffract X-rays marked the beginning of solid state physics as we know it today, and that technique remains one of the foremost tools of the field. The dynamical theory of diffraction was reviewed for X-rays by James (Vol. 15, 1963), and for electrons by Dederichs (Vol. 27, 1972). Neutron diffraction was discussed by one of its pioneers (Shull and Wollan, Vol. 2, 1956).

Guinier (Vol. 9, 1959) provides a comprehensive overview of how X-ray diffraction, and in particular small angle scattering, can be used to study compositional order and the formation of precipitates in solid solutions. An extensive update of this subject was given by Cohen (Vol. 39, 1986). Complementary information in such microstructural investigations is often obtained from transmission electron microscopy. Amelinckx (suppl. 6, 1964) provides a complete review of the technique and its applications to the study of interfaces and defects.

The emergence of synchrotron light sources has stimulated the growth of new techniques, such as the use of extended X-ray absorption fine structure for studying the coordination of specific chemical species (Hayes and Boyce, Vol. 37, 1982) or new applications of ultraviolet spectroscopy (Brown, Vol. 29, 1974).

Since the pioneering work of Bridgman, application of high pressure has become a valuable tool to determine equations of state of solids (Swenson, Vol. 11, 1960; Drickamer *et al.*, Vol. 19, 1966); to induce phase transformations (Bundy and Strong, Vol. 13, 1962); and to study the atomistic mechanisms of diffusion (Lazarus, Vol. 10, 1960). Complementary information can be obtained from transient experiments using shock waves (Rice *et al.*, Vol. 6, 1958; Doran and Linde, Vol. 19, 1966).

Extensive reviews in the series covered newly introduced techniques that included positron annihilation (Wallace, Vol. 10, 1960); cyclotron resonance (Lax and Mavroides, Vol. 11, 1960); electron spin resonance (Jarrett, Vol. 14, 1963); and optical modulation spectroscopy (Cardona, Suppl. 11, 1969). Articles on the rapidly expanding field of scanning probe microscopy will appear in upcoming volumes.

Structure of Condensed Matter

Knowing and understanding the structure of a material on the atomic scale is central to all of condensed matter physics. At the inception of the series, diffraction had revealed the structure of even the most complex crystals. Since then, however, a fascinating sequence of new structural problems has arisen, and new insights into old structural questions have been gained.

The structure of liquids, glasses, and quantum solids remains challenging. Ashcroft and Stroud (Vol. 33, 1978) review the thermodynamics and the correlations in simple liquid metals. Guyer (Vol. 23, 1969) discusses solid helium. The study of glasses is particularly useful, as the structural and thermal broadening of the correlations is less than in high temperature liquids. Diffraction studies on metallic glasses—the simplest kind—established the dense random packing of hard spheres as the structural model (Cargill, Vol. 30, 1975). The many tetrahedral configurations in this model supported F.C. Frank's early suggestion of an icosahedral paradigm

for the liquid structure, and inspired the polytetrahedral packing model, which is the closest we have come to a unified theory for the liquid structure (Nelson and Spaepen, Vol. 42, 1989; Yonezawa, Vol. 45, 1991).

A different approach was needed to understand the structure of directionally bonded amorphous materials, such as silicon or silica (Phillips, Vol. 37, 1982). Careful diffraction work and extensive modeling and simulation showed that the continuous random network, originally proposed by Zachariasen, is the model of choice (Wooten and Weaire, Vol. 40, 1987).

The discovery of quasi-crystals by Shechtman and coworkers, which was a serendipitous outgrowth of a study of rapidly solidified aluminum alloys, caused a revolution in crystallography: Like crystals, quasi-crystals have a polyhedral morphology and a sharp diffraction pattern, but because of their quasi-periodic structure they can exhibit symmetries, such as 5-fold rotational symmetry, that are prohibited in periodic structures (Nelson and Spaepen, Vol. 42, 1989). Quasi-periodicity is also key to the study of excitations in incommensurate crystal phases (Currat and Janssen, Vol. 41, 1988).

Bednorz and Müller's discovery of high-temperature superconductivity in La-Ba-Cu-O sparked the interest in the structure of cuprates, in particular the Y-B-C-O and its many derivatives (Beyers and Shaw, Vol. 42, 1989). Similarly, the synthesis of large quantities of fullerenes led to intensive structural investigation of their crystalline phases. (Axe, Moss, and Neumann, Vol. 48, 1994) De Heer *et al.*, (Vol. 40, 1987) review the structure of small metal clusters.

The structure of intercrystalline boundaries was elucidated greatly by the theoretical work of Bollmann and the experimental work of Balluffi and coworkers. An extensive review by Pond and Hirth (Vol. 47, 1994) shows the symmetries of these interfaces and how they can be modeled as arrays of defects. The structure of the crystal-melt interface remains a challenging problem. Although direct experimental structural evidence is lacking, modeling and indirect evidence from nucleation indicates that the liquid becomes increasingly ordered near the crystal (Spaepen, Vol. 47, 1994).

Electronic Structure

The power of the one-electron theory of solids, presaged by the work of Wigner and Seitz (1933; Vol. 1, 1955), was not appreciated until materials synthesis permitted the fabrication of crystals sufficiently perfect to permit cyclotron resonance (Lax and Mavroides, Vol. 11, 1960) and optical properties experiments in semiconductors (Cardona, Suppl. 11, 1969; Phillips, Vol. 18, 1966), and Fermi surface experiments in metals (Sellmyer, Vol. 33, 1978).

The methods for calculating band structures, like Herring's orthogo-

nalized plane wave method, the Korringa-Kohn-Rostoker method, Slater's augmented plane wave approach (Ziman, Vol. 26, 1971) were developed quite independently of experiment. The pseudo-potential approach in its early semiempirical form became particularly important, because the adjustment of relatively few parameters from experiment could be used to generate realistic and generally useful band structures (Cohen and Heine, Vol. 24, 1970). Volume 24, consisting of articles all authored or co-authored by Heine, in fact is devoted in its entirety to the explication of the pseudopotential ideas, their basis and their applications.

Standard band structure theories most frequently use k-space formalisms. However, as a result of increasing interest in transition metals during the 1960s and 1970s, it was noted that when, for example, d electrons (or f electrons in the rare earth metals) interact more strongly with their parent atoms than with neighbors, the electronic properties depend more on the local surroundings than on the long range periodicity. Volume 35 (1980), written by members of the Heine Group at Cambridge, is devoted to methods, such as Haydock's recursive solution of the Schrödinger equation, which led to the quantitative application of the tight binding or LCAO method (Heine, Vol. 35, 1980).

Group theory is an essential ingredient involving crystalline media having relevance to band structures or localized crystal or ligand field effects. Nussbaum (Vol. 18, 1966) provides an excellent pedagogical introduction, and Koster (Vol. 5, 1957) a definitive compilation of space groups and their representations. The relationship to crystal field theory is examined by Herzfeld and Meijer (Vol. 12, 1961).

Quasi-Particles

The band structure formally describes the energy spectrum of an additional electron or hole, interacting with the electrons already present. Because of its interaction with other electrons, the lifetime of a quasi-particle or elementary excitation, described by the imaginary part of the self energy, is finite. However, its spectral density is assumed to have a narrow Lorentzian resonance shape. Band theory generally neglects lifetime effects: The band structure defines a precise value of E for each value of k. Approximations to the energy shifts, described by the real part of the self-energy, associated with exchange and correlation effects, are, however, included in "first principles" calculations.

The relationship between many-body theory applied to quasi-particles and band structures is clarified by Hedin and Lundqvist (Vol. 23, 1969). They derive a local exchange-correlation potential term to be incorporated in the Schrödinger equation using a Green's function formalism.

The related density functional approach, due to Thomas and Fermi in

its simplest form, was formulated generally by Kohn, in collaboration with Hohenberg and Sham (Lang, Vol. 28, 1973; Callaway and March, Vol. 38, 1984). Because it is based on a variational treatment of an energy functional, it is best suited for the calculation of ground state properties of solids, surfaces, etc., using the local density approximation (LDA) (Kandel and Kaxiras, Vol. 54, 1999). The density functional formalism also leads to an exchange correlation potential similar to that derived by Hedin and Lunqvist (Vol. 23, 1969). The so-called GW method provides an improved quasi-particle picture, yielding, for example, correct semiconductor band gaps (Aulbur *et al.*, Vol. 54, 1999). However, the Coulomb interaction is treated to a relatively low order of approximation even in this case.

The electron-phonon interaction (Sham and Ziman, Vol. 15, 1963) gives rise to another kind of quasi-particle, the polaron (Appel, Vol. 21, 1968). In this case a single fermion (electron) interacts with a boson field (phonons). This problem attracted the interest even of those outside the solid state community. Feynman worked on this problem for two reasons: It is a nonrelativistic analog to quantum electrodynamics, and in strongly polar materials like the alkali or silver halides, the electron-phonon interaction is sufficiently strong to fall into the so-called intermediate coupling regime (Peeters and DeVreese, Vol. 38, 1984).

The quasi-particle picture can fail. In alloys, for example, in which the crystal symmetry is broken, the Bloch description for electrons may not be a good approximation. The particle spectral density, which is a narrow Lorentzian for a quasi-particle, assumes a very broad irregular shape in the presence of strong alloy scattering (Ehrenreich and Schwartz, Vol. 31, 1976).

Many-Body Effects

Many-body effects manifest themselves most directly in collective excitations, for example, plasmons (Pines, Vol. 1, 1955). More exotic forms of elementary excitations, such as helicons, are seen in plasmas in a magnetic field (Platzman and Wolff, Suppl. 13, 1973). The Kondo effect (Kondo; Heeger, Vol. 23, 1969), X-ray edge singularities (Mahan, Vol. 29, 1974), the specific heat of heavy fermion systems (Fulde, Keller, and Zwicknagl, Vol. 41, 1988), and the fractional quantum Hall effect (Isihara, Vol. 42, 1989) also depend explicitly on the proper treatment of Coulomb effects. The same is true for the binding energy of excitons (Knox, suppl. 5, 1963), the electron-hole liquid in semiconductors (Hensel, Phillips, and Thomas; Rice, Vol. 32, 1977) and the excitonic state at the semiconductor-semimetal transition (Halperin and Rice, Vol. 21, 1968).

White and Geballe have written a fine account of observed many-body effects including, among other topics, magnetism, ferroelectricity and magnetism in their book on *Long-Range Order in Solids* (suppl. 15, 1979).

Cohesion and Phase Stability

The understanding of the atomistic origins of the cohesion of condensed matter and the relative stability of the various phases has been, and remains, one of the central aims of solid state physics. It was one of the centerpieces of Seitz's classic 1940 work, *Modern Theory of Solids*, that motivated the origin of the series. Wigner and Seitz reviewed cohesion in metals in the first volume (1955). The introduction of the pseudo-potential described by Heine and Weaire (Vol. 24, 1970) was a major advance that greatly facilitated the calculation of specific problems of phase stability. The results of such calculations have been used, together with experimental data, to construct pair potentials, which are needed for calculations of complex configurations, such as surfaces or defects. Carlsson (Vol. 43, 1990) gives a detailed review of this approach, and how it can be extended to other energy functionals.

Tosi (Vol. 16, 1964) reviews the essentially classical problem of the stability of ionic crystals, the core of which requires methods to make the sum of long-range Coulomb potentials converge. Understanding the structure and stability of liquids, even simple monatomic metallic ones, remains one of the most challenging problems in condensed matter physics. Ashcroft and Stroud (Vol. 33, 1978) give an excellent overview of both the electronic and statistical mechanical aspects.

Application of electron theory to the prediction of phases in multicomponent systems goes back to Jones's zone theory (1934), which was an attempt to explain the remarkable empirical correlations between phase formation and the electron-atom ratio compiled by Hume-Rothery. The present state of phase prediction is highlighted in a number of recent articles. Pettifor (Vol. 40, 1987) examined the quantum mechanical origin of the heat of formation of alloys, and clarified the significance of the Miedema rules. De Fontaine (Vol. 34, 1979) contributed a comprehensive treatment of the theory of the atomic configurations in solid solutions, and elucidated understanding of the conditions for compositional ordering or clustering. The theory behind the complex phase diagrams in intercalated compounds was reviewed by Safran (Vol. 40, 1987). Yeomans (Vol. 41, 1988) provided a review of the use of the axial next nearest neighbor Ising model for calculating phase diagrams, and Martin and Bellon (Vol. 50, 1996) showed how phase equilibria shift as a result of sustained nonequilibrium conditions, such as irradiation or cold work.

Phase Transitions

The formation of a new phase from a matrix is a conceptually very rich phenomenon that has been approached from many disciplines: chemical kinetics, thermodynamics, statistical mechanics, and the solid state physics of phase stability discussed in the preceding.

Turnbull's initial overview of the thermodynamics and kinetics of phase changes (Vol. 3, 1956) is still one of the best tutorial introductions to the subject, and to Turnbull's main contribution: the theory of nucleation in condensed matter and its experimental verification. The subject of nucleation was revisited recently by Kelton (Vol. 45, 1991), who concentrated on the formation of crystal nuclei in liquids and glasses, and Wu (Vol. 50, 1996), who thoroughly reviewed classical nucleation theory, including transient nucleation. Spaepen (Vol. 47, 1994) showed what can be learned from nucleation about the structure of the crystal-melt interface, which is the barrier to the formation of the nuclei.

The stage beyond nucleation, crystal growth, is the subject of three articles. Pfann (Vol. 4, 1957) pioneered the technique of zone melting and refining, which was crucial in providing the ultraclean samples needed to make the development of the early semiconductor devices possible. Laudise and Nielsen (Vol. 12, 1961) were closely involved in the development of hydrothermal growth, essential for the production of quartz crystals that are ubiquitous in high technology devices. Parker (Vol. 25, 1970) provided an extensive general overview of the energetics and kinetics of the atomistic mechanisms that govern crystal growth.

Application of high pressure is a special tool to induce new phase transitions. Building on the pioneering work of Bridgman, Bundy and Strong (Vol. 13, 1962) studied the properties of materials at both high pressures and temperatures, in work that was instrumental in the development of synthetic diamond. Drickamer (Vol. 17, 1965) describes how the application of high pressure can produce band overlap and insulator-to-metal transitions. Samara and Peercy (Vol. 36, 1981) review soft-mode transitions at high pressure.

Order-disorder transitions manifest themselves in different systems. Best known are the Curie transition in ferromagnetic materials and the formation of ordered compounds in alloys. Ever since Ising proposed his model, it has been recognized that theoretical approaches are often broadly applicable across these different systems. This generality has been laid out elegantly and extensively in White and Geballe's volume on *Long-Range Order in Solids* (Suppl. 15, 1979). The phenomenology and early theories of order-disorder transitions in alloys were reviewed by Muto and Takagi (Vol. 1, 1955). Guttman (Vol. 3, 1956) provided a detailed discussion of the statistical mechanics of the problem. A comprehensive discussion of the modern approach, including methods such as the cluster variation method, can be found in de Fontaine's article (Vol. 47, 1994).

The study of the early stages of precipitation in metal alloys (the formation of Guinier-Preston zones) is a particularly fruitful area for sophisticated X-ray diffraction, as discussed by Guinier (Vol. 9, 1959) and Cohen (Vol. 39, 1986). The martensitic transformation is governed by an athermal displacive mechanism (as opposed to the diffusive mechanisms that govern most of

the other transitions already presented here), and its crystallography can be quite complex. Roitburd (Vol. 33, 1978) provided a review.

Electron Transport

Attention has been drawn to this topic since the nineteenth century, long before the solid state field was formally defined. According to Drude's theory, electrons should be scattered by ions constituting a metal. As a result, the calculated conductivity was far too small. This dilemma was not resolved until the advent of quantum mechanics and the realization that a perfect crystalline arrangement of atoms does not scatter electrons. Although still based on the Boltzmann equation, the study of transport properties became quantitative after the advent of zone refining and the fabrication of sufficiently pure semiconductors and metals. Moreover, scattering mechanisms were treated microscopically (Sham and Ziman, Vol. 15, 1963). Blatt's article (Vol. 4, 1957) mirrors this progress in a pedagogically oriented overview.

The discovery of the Gunn effect spurred the investigation of the transport properties of semiconductors at high electric fields (Conwell, suppl. 9, 1967). The development of the high magnetic fields necessary for determining Fermi surfaces also enabled the study of transport properties in this regime, some of which, like cyclotron resonance, themselves provided information concerning the electronic structure. Theoretical descriptions in this case were based on a Landau level picture, which included spin splittings (Kahn and Frederikse, Vol. 9, 1959; Kubo, Miyake, and Hashitsume, Vol. 17, 1965). In many cases quantum transport theory also needed to be invoked. This more general theory is central to descriptions of the transport properties of nanostructures (Beenakker and van Houten, Vol. 44, 1991) or the modeling of semiconductor devices such as resonant tunnel diodes (Ferry and Grubin, Vol. 49, 1995). The mathematical description of the latter strongly nonequilibrium case mandates the use of the so-called Keldysh formalism.

Tunneling phenomena, whose investigation was motivated in semiconductors by the discovery of the Esaki diode and in metals by Giaever's classic experiments involving $Al/Al_2O_3/$ Al sandwiches, have also been enormously productive as a spectroscopic tool in both normal and superconducting metals (Duke, Suppl. 10, 1969; Wolf, Vol. 30, 1975), and, as suggested in the preceding, in some device applications.

Lattice Dynamics and Phonon Transport

Because of the work of Einstein, Debye and Born, the theory of lattice vibrations in solids was firmly established before the discovery of quantum

16

mechanics. Indeed, highly influential solid state physics texts, such as that of Kittel, begin with this topic (and also group theory), whose basic ingredients can be described in classical terms, for pedagogical reasons. The authoritative book by Maradudin, *et al.*, concerning lattice dynamics in the harmonic approximation (Suppl. 3, second edition, 1971), begins with a description of the elements of the theory, and continues with the calculation of thermodynamic functions, X-ray and neutron scattering, and the effects of disorder and surfaces. A full theoretical description of anharmonic effects, responsible, for example, for the thermal conductivity of insulators, is provided by Leibfried and Ludwig (Vol. 12, 1961). Kwok (Vol. 20, 1967) describes the Green's function approach, which is useful not only for its discussion of intrinsically important subjects like the phonon-photon interaction, but also because it serves as a didactic introduction to this important theoretical technique. Applications to chalcopyrites and related semiconducting compounds and fullerenes are described by Miller, Mackinnon, and Weaire (Vol. 36, 1981) and Weaver and Poirier (Vol. 48, 1994), respectively.

The calculation of the thermal conductivity, for which phonon-phonon scattering poses a particularly difficult theoretical problem, is given detailed presentation by Klemens (Vol. 7, 1958). Mendelssohn and Rosenberg (Vol. 12, 1961) address metals at low temperatures, emphasizing the relative contribution of electrons and phonons to thermal conductivity. Slack (Vol. 34, 1979) gives a transparent model-oriented account applicable to nonmetallic crystals.

Structural Defects and Atomic Transport

One of the major achievements of materials science over the last fifty years has been the quantitative establishment of the presence and role of lattice imperfections. Traditionally, these imperfections are classified by their dimensionality: point defects (vacancies, interstitials, substitutional or interstitial impurities, and small clusters of these); line defects (dislocations, disclinations); and planar defects (stacking faults, grain boundaries).

The definitive experiment that established absolute vacancy concentrations was that of Simmons and Balluffi (1959), who measured, simultaneously, the temperature dependence of the atomic volume (from the lattice parameter) and the macroscopic volume of cubic crystals. To identify the difference between the fractional changes of these two quantities with the vacancy fraction, it is necessary to know that the relaxation of the atom positions around the vacancies causes a uniform elastic strain of the lattice without a shape change. This important result had been obtained by Eshelby, and is discussed in his article (Vol. 3, 1956) on the continuum theory of defects.

In most materials, atomic diffusion is governed by the formation and motion of point defects. Two articles describe the advances in the under-

standing of diffusion in metals. Lazarus (Vol. 10, 1960) discusses the basic thermodynamics and kinetics of the formation and motion of the point defects, including diffusion in binary alloys and the main experimental techniques. Peterson (Vol. 22, 1968) contributed a major update, which contains a clear discussion of the calculation of correlation effects, and a review of anomalous diffusion in body-centered cubic metals. Diffusion in ionic compounds introduces new complexities, such as charged defects and the effects of nonstoichiometry and foreign atoms. The classic analysis of these effects was developed by Kröger and Vink, and is extensively reviewed in volume 3 (1956).

The development of nuclear physics and its applications (reactors, accelerators) raised new materials problems. Energetic particles can displace the atoms from their equilibrium positions, and create point defects in concentrations far beyond the equilibrium values. The seminal review of this field was by Seitz and Koehler (Vol. 2, 1956); recently Averback and Diaz de la Rubia (Vol. 51, 1997) contributed an update that goes beyond the original single particle analysis and reviews the insights that can be obtained from computer simulation. Martin and Bellon (Vol. 50, 1996) reviewed the changes in phase equilibria produced under these strongly nonequilibrium conditions.

The conceptual and experimental discovery of the dislocation was central to understanding the plastic deformation of crystals. The basic dislocation theory, rooted in continuum mechanics, can be found in the articles by Eshelby (Vol. 3, 1956) and de Wit (Vol. 10, 1960). Gilman and Johnston (Vol. 13, 1962) performed the pioneering experiments on the visualization and mobility of dislocations. The bulk of the experimental evidence for the static and dynamic behavior of dislocations came from transmission electron microscopy, as reviewed extensively by Amelinckx (suppl. 6, 1964). Another line defect, the disclination, has been invoked to understand the structure of both simple glasses and polytetrahedral packings (Nelson and Spaepen, Vol. 42, 1989).

Arrays of dislocations form low-angle grain boundaries; the early evidence for these is reviewed by Amelinckx and Dekeyser (1959). A thorough discussion of the sophisticated geometry underlying coherent and incoherent interfaces and the associated interfacial defects can be found in the article by Pond and Hirth (Vol. 47, 1994). The curvature-driven motion of grain boundaries in a polycrystal lowers the interfacial area of the system. The resulting grain growth is a complex multibody phenomenon, and is reviewed by Weaire and McMurry (Vol. 50, 1996).

Mechanical Properties

The study of the mechanical behavior of materials is wide-ranging in its approach: from the atomistic to the continuum level; from fundamental

to applied. The elastic behavior of crystals—the instantaneous, reversible response of the lattice to stress—was first reviewed by Huntington (Vol. 7, 1958), whose article remains a valuable collection of data as well. Wallace (Vol. 25, 1970) extended this to include the nonlinear part of the response. Keyes (Vol. 20, 1967) reviewed the electronic basis of the elastic behavior of semiconductors.

Plastic deformation—the time-dependent and irreversible response—is governed often by the creation and motion of dislocations, the theoretical and experimental aspects of which we touched on in the preceding section on structure defects and atomic transport (Eshelby, de Wit, Gilman and Johnston, Amelinckx). The plastic deformation of crystals with the diamond structure (in particular, silicon and germanium) is not only of great technological importance, but is in important ways different from that in close-packed metallic crystals: The dislocations are narrower, have characteristic orientations and cores, and are generally less mobile. Alexander and Haasen (Vol. 22, 1968) wrote a classic review of the subject.

Thomson contributed two articles about the continuum mechanics and atomistics of fracture. The first (Vol. 39, 1986) is a basic introduction to the subject, in which he treats the elastic stress fields around cracks, the atomistics of crack tips and the role of dislocations. The second article, with Carlsson (Vol. 51, 1997), focuses on atomistic calculations of the behavior at the crack tip and on how to link the results with the behavior at larger length scales. Evans and Zok's article (Vol. 47, 1994) shows how the sophisticated combination of continuum mechanics and microstructural knowledge can be used to optimize the mechanical behavior (in particular, fracture) of composite materials.

Optical Properties

Atomic spectroscopy has had enormous influence in the development of quantum mechanics. Optical spectroscopy over an extended frequency range, extending beyond the far ultraviolet, has played a similar role in the understanding of the band structure of solids. The measurement of reflectance, together with a Kramers-Kronig analysis of the data, provides both real and imaginary parts of the dielectric functions, whose structure can in turn be associated with strong interband transitions ranging beyond the fundamental gap in semiconductors.

Cardona's optical modulation spectroscopy, described in a supplementary volume bearing that title (Suppl. 11, 1969), is important because of its ability to determine spectroscopic signatures associated with band structure features more precisely by examining derivatives obtained by modulating the photon wavelength or stress and electric fields applied to samples that amplify these small changes enormously.

Stern (Vol. 15, 1963) presents a very approachable elementary account of some of these matters. Brown (Vol. 29, 1974) discusses the highly important ultraviolet spectroscopy of solids utilizing synchrotron radiation. Phillips (Vol. 18, 1966) and Nilsson (Vol. 29, 1974) survey the then existing data and its interpretation for semiconductors and metals, respectively.

Bloch Electrons in Strong External Fields

The quantum mechanical and dynamical description of Bloch electrons in the presence of, for example, strong electric and magnetic fields, which give rise respectively to Stark ladders and Landau levels, requires band theoretic formalisms such as the crystal momentum and coordinate representations (Blount, Vol. 13, 1962), or Zak's kq-representation (Vol. 27, 1972). Bands can be strongly mixed under these circumstances. This mixing is responsible for Zener tunneling and magnetic breakdown. Group theoretical approaches based on vector and ray representations and magnetic translation groups also provide important insight (Brown, Vol. 22, 1968).

Semiconductors

In current solid state research, semiconductors serve either as model systems to explore qualitatively new physical effects, for example, lower dimensional systems such as nanostructures (Beenakker and van Houten, Vol. 44, 1991) and the quantum Hall effect (Isihara, Vol. 42, 1989), or as device materials, whose physical functionality requires basic physical understanding. In the latter category, there are early articles on the lead salts (Dalven, Vol. 28, 1973), which served as infrared detectors, luminescence (Klick and Schulman, Vol. 5, 1957), electroluminescence (Piper and Williams, Vol. 6, 1958), important in fluorescent bulbs and TV displays, and the classic article by Welker and Weiss on III-V compounds (Vol. 3, 1956), which describes the research that anticipated the possible device applications of these materials.

Of more recent interest is work concerning II-VI visible light emitters (Nurmikko and Gunshor, Vol. 49, 1995), light emission from silicon (Kimerling *et al.*, Vol. 50, 1996), semiconductor device physics of conjugated polymers (Greenham and Friend, Vol. 49, 1995), "good" thermoelectrics (Mahan, Vol. 51, 1997), electromigration (Sorbello, Vol. 51, 1997), modeling of quantum transport in semiconductor devices (Ferry and Grubin, Vol. 49, 1995), and semiconductor heterostructures (Bastard, Brum, and Ferreira, Vol. 44, 1991) together with their band offsets (Yu *et al.*, Vol. 46, 1992).

Magnetism

Because of their great potential importance in magnetic recording (Bertram and Zhu, Vol. 46, 1992), giant and colossal magnetoresistance effects have caused a pronounced resurgence in magnetism research (Levy, Vol. 47, 1994).

Many of the earlier articles on magnetism that have appeared in the series remain classics. Anderson's account of the theory of superexchange in insulators and semiconductors (Vol. 14, 1963) is still authoritative. So are the articles by Kittel concerning indirect exchange interactions in metals (Vol. 22, 1968) and ferromagnetic domain theory (with Galt, Vol. 3, 1956).

There are also definitive articles and supplements describing nuclear magnetic resonance (Pake, Vol. 2, 1956), electron paramagnetism (e.g., Knight, Vol. 2, 1956; Low, suppl. 2, 1960; Jarrett, Vol. 14, 1963), localized magnetic moments (Kondo, Vol. 23, 1969; Heeger, Vol. 23, 1969), neutron diffraction (Shull and Wollan, Vol. 2, 1956), relaxation phenomena (Yafet, Vol. 14, 1963; Hebel, Vol. 15, 1963), magnetic properties of rare earth metals (Cooper, Vol. 21, 1968), and heavy fermion systems (Fulde, Keller, and Zwicknagl, Vol. 41, 1988).

Simulations

Simulations of complicated physical phenomena are becoming ever more important with rapidly increasing computer speeds and capacities. Indeed, one might say that experimental and theoretical physics are now being supplemented by a third area, computational physics, which shares ingredients with the other two, while at the same time being substantially different. Several articles have appeared in the series since the late-1980s that explore several areas addressed by this approach. Yonezawa (Vol. 45, 1991) summarizes her work concerning simulated glass transitions and relaxation in disordered systems; de Fontaine (Vol. 47,1994) describes the cluster approach to order-disorder transformation in alloys; Averback and Diaz de la Rubia (Vol. 51, 1997) review displacement damage in irradiated metals and semiconductors; Ferry and Grubin (Vol. 49, 1995) show how to model quantum transport in semiconductor devices. Most recently, Kandel and Kaxiras (Vol. 54, 1999) describe their first-principles calculations concerning semiconductor thin film growth. Evidently, the Solid State Physics series will emphasize computational physics to a far greater extent in the future.

Other Topics: The Future

Although the *Solid State Physics* series has thus far not been as extensively concerned with superconductivity or ferroelectricity, even though such atten-

tion is entirely warranted, there are a number of important contributions. The book by White and Geballe (Suppl. 15, 1979), *Long Range Order in Solids*, provides an excellent account of both areas. Allen and Mitrović (Vol. 37, 1982) discuss the theory of the superconducting transition temperature in the strong coupling regime. An entire volume (Vol. 42, 1989), published shortly after their discovery, is devoted to the "new" high-temperature superconductors (Tinkham and Lobb, Vol. 42, 1989; and others). The subject will receive increased emphasis in forthcoming volumes (Millis, in preparation). The classic early review of ferroelectricity by Känzig (Vol. 4, 1957) will also be updated soon (Newnham, in preparation).

There are also articles describing various aspects of surface science. Davison and Levine (Vol. 25, 1970) discuss surface states and Lang (Vol. 28, 1973) the electronic structure of metal surfaces using the then recently developed density-functional formalism. Webb and Lagally (Vol. 28, 1973) summarize their work on low energy elastic scattering. Gomer reviews both theoretical descriptions and experimental results pertaining to chemisorption on metals (Vol. 30, 1975) and Osgood and Wang (Vol. 51, 1997) discuss image states. This is a rapidly growing area, driven by the multiplying applications of scanning probe techniques, which will be covered more extensively in the future.

The same is true for organic materials. Recent contributions have included an important article concerning semiconductor device physics of conjugated polymers (Greenham and Friend, Vol. 49, 1995), and an entire volume (Vol. 48, 1994) on fullerenes, including the structure and dynamics (Axe, Moss, and Neumann), doped superconductors (Lieber and Zhang), materials preparation (Lieber and Chen), and electronic, lattice, and other physical properties (Pickett; Weaver and Poirier). Liquid crystal physics is treated extensively in a book edited by Liebert (Suppl. 14, 1978). It is written by eminent contributors, including de Gennes. It is clear that the series must emphasize such new materials to a far greater extent as condensed matter physics becomes increasingly "soft."

III. Contents: Subject Index

III. Contents: Subject Index

1. Materials Synthesis 26
2. Experimental Techniques 26
3. Structure of Condensed Matter 27
4. Band Theory 28
5. Electronic Structure 29
6. Group Theory 31
7. Elementary Excitations 31
8. Many-Body Effects 31
9. Cohesion and Phase Stability 32
10. Phase Transitions 33
11. Electron Transport 34
12. Lattice Dynamics 35
13. Phonon Transport 36
14. Structural Defects and Atomic Transport 36
15. Mechanical Properties 37
16. Surfaces & Interfaces 38
17. Disordered Systems 39
18. Optical Properties 39
19. Bloch Electrons in Strong External Fields 41
20. Semiconductors 41
21. Magnetism 42
22. Ferroelectrics and Dielectrics 44
23. Superconductivity 44
24. Alkali Halides 44
25. Transition Metals & Oxides 45
26. Alloys 45
27. Organic Materials 45
28. Rare Earth Metals 46
29. Simulations 47
30. Data Collections 47

1. Materials Synthesis

Pfann, W.G.: Techniques of Zone Melting and Crystal Growing, **4,** 423 (1957)

Laudise, R.A., and Nielsen, J.W.: Hydrothermal Crystal Growth, **12,** 149 (1961)

Bundy, F.P., and Strong, H.M.: Behavior of Metals at High Temperatures and Pressures, **13,** 81 (1962)

Parker, R.L.: Crystal Growth Mechanisms: Energetics, Kinetics, and Transport, **25,** 151 (1970)

Keller, P., and Liebert, L.: Liquid-Crystal Synthesis for Physicists, *in* Supplement 14–Liquid Crystals, 19 (1978)

de Heer, Walt A., Knight, W.D., Chou, M.Y., and Cohen, Marvin L.: Electronic Shell Structure and Metal Clusters, **40,** 93 (1987)

Lieber, Charles, M., and Chen, Chia-Chun: Preparation of Fullerenes and Fullerene-Based Materials, **48,** 109 (1994)

2. Experimental Techniques

Shull, C.G., and Wollan, E.O.: Applications of Neutron Diffraction to Solid State Problems, **2,** 137 (1956)

Rice, M.H., McQueen, R.G., and Walsh, J.M.: Compression of Solids by Strong Shock Waves, **6,** 1 (1958)

Becker, J.A.: Study of Surfaces by Using New Tools, **7,** 379 (1958)

Kothari, L.S., and Singwi, K.S.: Interaction of Thermal Neutrons with Solids, **8,** 109 (1959)

Heer, Ernst, and Novey, Theodore B.: The Interdependence of Solid State Physics and Angular Distribution of Nuclear Radiations, **9,** 199 (1959)

Guinier, Andre: Heterogeneities in Solid Solutions, **9,** 293 (1959)

Wallace, Philip R.: Positron Annihilation in Solids and Liquids, **10,** 1 (1960)

Lazarus, David: Diffusion in Metals, **10,** 71 (1960)

Swenson, C.A.: Physics at High Pressure, **11,** 41 (1960)

Bube, Richard H.: Imperfection Ionization Energies in CdS-Type Materials by Photoelectronic Techniques, **11,** 223 (1960)

Lax, Benjamin, and Mavroides, John G.: Cyclotron Resonance, **11,** 261 (1960)

Bundy, F.P., and Strong, H.M.: Behavior of Metals at High Temperatures and Pressures, **13,** 81 (1962)

Jarrett, H.S.: Electron Spin Resonance Spectroscopy in Molecular Solids, **14,** 215 (1963)

James, R.W.: The Dynamical Theory of X-Ray Diffraction, **15,** 53 (1963)

Amelinckx, S., Supplement 6–The Direct Observation of Dislocations (1964)

Einspruch, Norman, G.: Ultrasonic Effects in Semiconductors, **17,** 217 (1965)

Drickamer, H.G., Lynch, R.W., Clendenen, R.L., and Perez-Albuerne, E.A.: X-Ray Diffraction Studies of the Lattice Parameters of Solids under Very High Pressure, **19,** 135 (1966)

Doran, Donald G., and Linde, Ronald K.: Shock Effects in Solids, **19,** 229 (1966)

Brill, R.: Determination of Electron Distribution in Crystals by Means of X-Rays, **20,** 1 (1967)

Cardona, Manuel: Supplement 11–Optical Modulation Spectroscopy of Solids (1969)

Dederichs, P.H.: Dynamical Diffraction Theory by Optical Potential Methods, **27,** 135 (1972)

Brown, Frederick C.: Ultraviolet Spectroscopy of Solids with the Use of Synchrotron Radiation, **29,** 1 (1974)

Gomer, Robert: Chemisorption on Metals, **30,** 93 (1975)

Sellmyer, D.J.: Electronic Structure of Metallic Compounds and Alloys: Experimental Aspects, **33,** 83 (1978)

Schnatterly, S.E.: Inelastic Electron Scattering Spectroscopy, **34,** 275 (1979)

Hayes, T.M., and Boyce, J.B.: Extended X-Ray Absorption Fine Structure Spectroscopy, **37,** 173 (1982)

Cohen, Jerome B.: The Internal Structure of Guinier-Preston Zones in Alloys, **39,** 131 (1986)

McGreevy, Robert L.: Experimental Studies of the Structure and Dynamics of Molten Alkali and Alkaline Earth Halides, **40,** 247 (1987)

3. Structure of Condensed Matter

Wells, A.F.: The Structures of Crystals, **7,** 425 (1958)

Drickamer, H.G., Lynch, R.W., Clendenen, R.L., and Perez-Albuerne, E.A.: X-Ray Diffraction Studies of the Lattice Parameters of Solids under Very High Pressure, **19,** 135 (1966)

Cargill, G. S., III: Structure of Metallic Alloy Glasses, **30,** 227 (1975)

Ashcroft, N.W., and Stroud, D.: Theory of the Thermodynamics of Simple Liquid Metals, **33,** 1 (1978)

Charvolin, Jean, and Tardieu, Annette: Lyotropic Liquid Crystals: Structures and Molecular Motions, *in* Supplement 14–Liquid Crystals, 209 (1978)

Phillips, J.C.: Spectroscopic and Morphological Structure of Tetrahedral Oxide Glasses, **37,** 93 (1982)

Wooten, F., and Weaire, D.: Modeling Tetrahedrally Bonded Random Networks by Computer, **40,** 1 (1987)

de Heer, Walt A., Knight, W.D., Chou, M.Y., and Cohen, Marvin L.: Electronic Shell Structure and Metal Clusters, **40,** 93 (1987)

McGreevy, Robert L.: Experimental Studies of the Structure and Dynamics of Molten Alkali and Alkaline Earth Halides, **40,** 247 (1987)

Currat, R., and Janssen, T.: Excitations in Incommensurate Crystal Phases, **41,** 201 (1988)

Nelson, David R., and Spaepen, Frans: Polytetrahedral Order in Condensed Matter, **42,** 1 (1989)

Beyers, R., and Shaw, T.M.: The Structure of $Y_1Ba_2Cu_3O_{7-\delta}$ and Its Derivatives, **42,** 135 (1989)

Sood, Ajay K.: Structural Ordering in Colloidal Suspensions, **45,** 1 (1991)

Yonezawa, Fumiko: Glass Transition and Relaxation of Disordered Structures, **45,** 179 (1991)

Spaepen, Frans: Homogeneous Nucleation and the Temperature Dependence of the Crystal-Melt Interfacial Tension, **47,** 1 (1994)

Pond, R.C., and Hirth, J.P.: Defects at Surfaces and Interfaces, **47,** 287 (1994)

Axe, J.D., Moss, S.C., and Neumann, D.A.: Structure and Dynamics of Crystalline C_{60}, **48,** 149 (1994)

4. Band Theory

Reitz, John R.: Methods of the One-Electron Theory of Solids, **1,** 1 (1955)

Ham, Frank S.: The Quantum Defect Method, **1,** 127 (1955)

Woodruff, Truman O.: The Orthogonalized Plane-Wave Method, **4,** 367 (1957)

Callaway, Joseph: Electron Energy Bands in Solids, **7,** 99 (1958)

Blount, E.I.: Formalisms of Band Theory, **13,** 306 (1962)

Nussbaum, Allen: Crystal Symmetry, Group Theory, and Band Structure Calculations, **18,** 165 (1966)

Heine, Volker: The Pseudopotential Concept, **24,** 1 (1970)

Cohen, Marvin L., and Heine, Volker: The Fitting of Pseudopotentials to Experimental Data and Their Subsequent Application, **24,** 37 (1970)

Heine, Volker, and Weaire, D.: Pseudopotential Theory of Cohesion and Structure, **24,** 249 (1970)

Ziman, J.M.: The Calculation of Bloch Functions, **26,** 1 (1971)

Dimmock, J.O.: The Calculation of Electronic Energy Bands by the Augmented Plane Wave Method, **26,** 103 (1971)

Dederichs, P.H.: Dynamical Diffraction Theory by Optical Potential Methods, **27,** 135 (1972)

Heine, Volker: Electronic Structure from the Point of View of the Local Atomic Environment, **35,** 1 (1980)

Bullett, D.W.: The Renaissance and Quantitative Development of the Tight-Binding Method, **35,** 129 (1980)

Haydock, Roger: The Recursive Solution of the Schrödinger Equation, **35,** 215 (1980)

Kelly, M.J.: Applications of the Recursion Method to the Electronic Structure from an Atomic Point of View, **35,** 295 (1980)

Miller, A., MacKinnon, A., and Weaire, D.: Beyond the Binaries—The Chalcopyrite and Related Semiconducting Compounds, **36,** 119 (1981)

Hui, P.M., and Johnson, Neil F.: Photonic Band-Gap Materials, **49,**151 (1995)

5. Electronic Structure

Wigner, Eugene P., and Seitz, Frederick: Qualitative Analysis of the Cohesion in Metals, **1,** 97 (1955)

Ham, Frank S.: The Quantum Defect Method, **1,** 127 (1955)

Koster, G.F.: Space Groups and Their Representations, **5,** 174 (1957)

Kohn, W.: Shallow Impurity States in Silicon and Germanium, **5,** 257 (1957)

Callaway, Joseph: Electron Bands in Solids, **7,** 99 (1958)

Lewis, H.W.: Wave Packets and Transport of Electrons in Metals, **7,** 353 (1958)

Wolf, H.C.: The Electronic Spectra of Aromatic Molecular Crystals, **9,** 1 (1959)

Gourary, Barry S., and Adrian, Frank J.: Wave Functions for Electron-Excess Color Centers in Alkali Halide Crystals, **10,** 127 (1960)

Lax, Benjamin, and Mavroides, John G.: Cyclotron Resonance, **11,** 261 (1960)

Herzfeld, Charles M., and Meijer, Paul H.E.: Group Theory and Crystal Field Theory, **12,** 1 (1961)

Blount, E.I.: Formalisms of Band Theory, **13,** 306 (1962)

Jørgensen, Klixbüll, Chr.: Chemical Bonding Inferred from Visible and Ultraviolet Absorption Spectra, **13,** 375 (1962)

Drickamer, H.G.: The Effect of High Pressure on the Electronic Structure of Solids, **17,** 1 (1965)

Phillips, J.C.: The Fundamental Optical Spectra of Solids, **18,** 55 (1966)

Nussbaum, Allen: Crystal Symmetry, Group Theory, and Band Structure Calculations, **18,** 165 (1966)

Cardona, Manuel: Supplement 11–Optical Modulation Spectroscopy of Solids (1969)

Cohen, Marvin L., and Heine, Volker: The Fitting of Pseudo-potentials to Experimental Data and Their Subsequent Application, **24,** 37 (1970)

Davison, S.G., and Levine, J.D.: Surface States, **25,** 1 (1970)

Ziman, J.M.: The Calculation of Bloch Functions, **26,** 1 (1971)

Abrikosov, A.A.: Supplement 12–Introduction to the Theory of Normal Metals (1972)

Dalven, Richard: Electronic Structure of PbS, PbSe, and PbTe, **28,** 179 (1973)

Lang, Norton D.: The Density-Functional Formalism and the Electronic Structure of Metal Surfaces, **28,** 225 (1973)

Busch, G., and Güntherodt, H.-J.: Electronic Properties of Liquid Metals and Alloys, **29,** 235 (1974)

Joannopoulos, J.D., and Cohen, Marvin, L.: Theory of Short-Range Order and Disorder in Tetrahedrally Bonded Semiconductors, **31,** 71 (1976)

Ehrenreich, H., and Schwartz, L. M.: The Electronic Structure of Alloys, **31,** 149 (1976)

Sellmyer, D.J.: Electronic Structure of Metallic Compounds and Alloys: Experimental Aspects, **33,** 83 (1978)

Heine, Volker: Electronic Structure from the Point of View of the Local Atomic Environment, **35,** 1 (1980)

Bullett, D.W.: The Renaissance and Quantitative Development of the Tight-Binding Method, **35,** 129 (1980)

Haydock, Roger: The Recursive Solution of the Schrödinger Equation, **35,** 215 (1980)

Kelly, M.J.: Applications of the Recursion Method to the Electronic Structure from an Atomic Point of View, **35,** 295 (1980)

Miller, A., MacKinnon, A., and Weaire, D.: Beyond the Binaries— The Chalcopyrite and Related Semiconducting Compounds, **36,** 119 (1981)

Callaway, J., and March, N.H.: Density Functional Methods: Theory and Applications, **38,** 135 (1984)

Zunger, Alex: Electronic Structure of $3d$ Transition-Atom Impurities in Semiconductors, **39,** 275 (1986)

Pettifor, D.G.: A Quantum-Mechanical Critique of the Miedema Rules for Alloy Formation, **40,** 43 (1987)

de Heer, Walt A., Knight, W.D., Chou, M.Y., and Cohen, Marvin L.: Electronic Shell Structure and Metal Clusters, **40,** 93 (1987)

Fulde, Peter, Keller, Joachim, and Zwicknagl, Gertrud: Theory of Heavy Fermion Systems, **41,** 1 (1988)

Hass, K.C.: Electronic Structure of Copper-Oxide Superconductors, **42,** 213 (1989)

Rasolt, M.: Continuous Symmetries and Broken Symmetries m Multivalley Semiconductors and Semimetals, **43,** 93 (1990)

Bastard, G., Brum, J.A., and Ferreira, R.: Electronic States in Semiconductor Heterostructures, **44,** 229 (1991)

Pickett, Warren E.: Electrons and Phonons in C_{60}-Based Materials, **48,** 225 (1994)

Greenham, Neil C., and Friend, Richard H.: Semiconductor Device Physics of Conjugated Polymers, **49,** 1 (1995)

Osgood, Richard M. Jr., and Wang, Xiaoyi: Image States on Single-Crystal Metal Surface, **51,** 1 (1997)

6. Group Theory

Koster, G.F.: Space Groups and Their Representations, **5,** 174 (1957)

Smith, Charles S.: Macroscopic Symmetry and Properties of Crystals, **6,** 175 (1958)

Herzfeld, Charles M., and Meijer, Paul H.E.: Group Theory and Crystal Field Theory, **12,** 1 (1961)

Nussbaum, Allen: Crystal Symmetry, Group Theory, and Band Structure Calculations, **18,** 165 (1966)

Brown, E.: Aspects of Group Theory in Electron Dynamics, **22,** 313 (1968)

7. Elementary Excitations

Pines, David: Electron Interaction in Metals, **1,** 367 (1955)

Knox, Robert S.: Supplement 5—Theory of Excitons (1963)

Sham, L.J., and Ziman, J.M.: The Electron-Phonon Interaction, **15,** 221 (1963)

Sturge, M.D.: The Jahn-Teller Effect in Solids, **20,** 91 (1967)

Appel, J.: Polarons, **21,** 193 (1968)

Hedin, Lars, and Lundqvist, Stig: Effects of Electron-Electron and Electron-Phonon Interactions on the One-Electron States of Solids, **23,** 1 (1969)

Glicksman, Maurice: Plasmas in Solids, **26,** 275 (1971)

Ehrenreich, H., and Schwartz, L. M.: The Electronic Structure of Alloys, **31,** 149 (1976)

Schnatterly, S.E.: Inelastic Electron Scattering Spectroscopy, **34,** 275 (1979)

Peeters, F.M., and Devreese, J.T.: Theory of Polaron Mobility, **38,** 81 (1984)

Singh, Jai: The Dynamics of Excitons, **38,** 295 (1984)

Currat, R., and Janssen, T.: Excitations in Incommensurate Crystal Phases, **41,** 201 (1988)

Echenique, P.M., Flores, F., and Ritchie, R.H.: Dynamic Screening of Ions in Condensed Matter, **43,** 229 (1990)

Aulbur, W.G., Jönssen, L., and Wilkins, J.W.: Quasiparticle Calculations in Solids, (**54,** 1999; to be published)

8. Many-Body Effects

Pines, David: Electron Interaction in Metals, **1,** 367 (1955)

Knox, Robert S.: Supplement 5—Theory of Excitons (1963)

Sham, L.J., and Ziman, J.M.: The Electron-Phonon Interaction, **15,** 221 (1963)

Halperin, B.I., and Rice, T.M.: The Excitonic State at the Semiconductor-Semimetal Transition, **21,** 115 (1968)

Hedin, Lars, and Lundqvist, Stig: Effects of Electron-Electron and Electron-Phonon Interactions on the One-Electron States of Solids, **23,** 1 (1969)

Kondo, J.: Theory of Dilute Magnetic Alloys, **23,** 183 (1969)

Guyer, R.A.: The Physics of Quantum Crystals, **23,** 413 (1969)

Glicksman, Maurice: Plasmas in Solids, **26,** 275 (1971)

Lang, Norton D.: The Density-Functional Formalism and the Electronic Structure of Metal Surfaces, **28,** 225 (1973)

Platzman, P.M., and Wolff, P.A.: Supplement 13—Waves and Interactions in Solid State Plasmas (1973)

Mahan, G.D.: Many-Body Effects on X-Ray Spectra of Metals, **29,** 75 (1974)

Wolf, E.L.: Nonsuperconducting Electron Tunneling Spectroscopy, **30,** 1 (1975)

Rice, T.M.: The Electron-Hole Liquid in Semiconductors: Theoretical Aspects, **32,** 1 (1977)

Hensel, J.C., Phillips, T.G., and Thomas, G.A.: The Electron-Hole Liquid in Semiconductors: Experimental Aspects, **32,** 87 (1977)

White, Robert M., and Geballe, Theodore H.: Supplement 15—Long Range Order in Solids (1979)

Singwi, K.S., and Tosi, M.P.: Correlations in Electron Liquids, **36,** 177 (1981)

Fulde, Peter, Keller, Joachim, and Zwicknagl, Gertrud: Theory of Heavy Fermion Systems, **41,** 1 (1988)

Isihara, A.: Electron Correlations in Two Dimensions, **42,** 271 (1989)

Echenique, P.M., Flores, F., and Ritchie, R.H.: Dynamic Screening of Ions in Condensed Matter, **43,** 229 (1990)

9. Cohesion and Phase Stability

Wigner, Eugene P., and Seitz, Frederick: Qualitative Analysis of the Cohesion in Metals, **1,** 97 (1955)

Borelius, G.: The Changes in Energy Content, Volume, and Resistivity with Temperature in Simple Solids and Liquids, **15,** 1 (1963)

Tosi, Mario P.: Cohesion of Ionic Solids in the Born Model, **16,** 1 (1964)

Heine, Volker, and Weaire, D.: Pseudopotential Theory of Cohesion and Structure, **24,** 249 (1970)

Ashcroft, N.W., and Stroud, D.: Theory of the Thermodynamics of Simple Liquid Metals, **33,** 1 (1978)

de Fontaine, D.: Configurational Thermodynamics of Solid Solutions, **34,** 73 (1979)

Wooten, F., and Weaire, D.: Modeling Tetrahedrally Bonded Random Networks by Computer, **40,** 1 (1987)

Pettifor, D.G.: A Quantum-Mechanical Critique of the Miedema Rules for Alloy Formation, **40,** 43 (1987)

Safran, S.A.: Stage Ordering in Intercalation Compounds, **40,** 183 (1987)

Yeomans, Julia: The Theory and Application of Axial Ising Models, **41,** 151 (1988)

Carlsson, A.E.: Beyond Pair Potentials in Elemental Transition Metals and Semiconductors, **43,** 1 (1990)

Martin, Georges, and Bellon, Pascal: Driven Alloys, **50,** 189 (1996)

10. Phase Transitions

Muto, Toshinosuke, and Takagi, Yutaka: The Theory of Order-Disorder Transitions in Alloys, **1,** 194 (1955)

Guttman, Lester: Order-Disorder Phenomena in Metals, **3,** 145 (1956)

Turnbull, David: Phase Changes, **3,** 225 (1956)

Pfann, W.G.: Techniques of Zone Melting and Crystal Growing, **4,** 423 (1957)

Borelius, G.: Changes of State of Simple Solid and Liquid Metals, **6,** 65 (1958)

Guinier, Andre: Heterogeneities in Solid Solutions, **9,** 293 (1959)

Laudise, R.A., and Nielsen, J.W.: Hydrothermal Crystal Growth, **12,** 149 (1961)

Bundy, F.P., and Strong, H.M.: Behavior of Metals at High Temperatures and Pressures, **13,** 81 (1962)

Drickamer, H.G.: The Effect of High Pressure on the Electronic Structure of Solids, **17,** 1 (1965)

Adler, David: Insulating and Metallic States in Transition Metal Oxides, **21,** 1 (1968)

Halperin, B.I., and Rice, T.M.: The Excitonic State at the Semiconductor-Semimetal Transition, **21,** 115 (1968)

Parker, R.L.: Crystal Growth Mechanisms: Energetics, Kinetics, and Transport, **25,** 151 (1970)

Weger, M., and Goldberg, I.B.: Some Lattice and Electronic Properties of the β-Tungstens, **28,** 1 (1973)

Charvolin, Jean, and Tardieu, Annette: Lyotropic Liquid Crystals: Structures and Molecular Motions, *in* Supplement 14—Liquid Crystals, 209 (1978)

Roitburd, A.L.: Martensitic Transformation as a Typical Phase Transformation in Solids, **33,** 317 (1978)

White, Robert M., and Geballe, Theodore H.: Supplement 15—Long Range Order in Solids (1979)

Samara, G.A., and Peercy, P.S.: The Study of Soft-Mode Transitions at High Pressure, **36,** 1 (1981)

Cohen, Jerome B.: The Internal Structure of Guinier-Preston Zones in Alloys, **39,** 131 (1986)

Nelson, David R., and Spaepen, Frans: Polytetrahedral Order in Condensed Matter, **42,** 1 (1989)

Kelton, K.F.: Crystal Nucleation in Liquids and Glasses, **45,** 75 (1991)

Spaepen, Frans: Homogeneous Nucleation and the Temperature Dependence of the Crystal-Melt Interfacial Tension, **47,** 1 (1994)

de Fontaine, D.: Cluster Approach to Order-Disorder Transformations in Alloys, **47,** 33 (1994)

Wu, David T.: Nucleation Theory, **50,** 37 (1996)

Martin, Georges, and Bellon, Pascal: Driven Alloys, **50,** 189 (1996)

Averback, R.S., and Diaz de la Rubia, T.: Displacement Damage in Irradiated Metals and Semiconductors, **51,** 281 (1997)

11. Electron Transport

Blatt, Frank J.: Theory of Mobility of Electrons in Solids, **4,** 200 (1957)

Jan, J.-P.: Galvanomagnetic and Thermomagnetic Effects in Metals, **5,** 1 (1957)

Dekker, A.J.: Secondary Electron Emission, **6,** 251 (1958)

Heiland, G., Mollwo, E., and Stöckmann, F.: Electronic Processes in Zinc Oxide, **8,** 191 (1959)

Scanlon, W.W.: Polar Semiconductors, **9,** 83 (1959)

Kahn, A.H., and Frederikse, H.P.R.: Oscillatory Behavior of Magnetic Susceptibility and Electronic Conductivity, **9,** 257 (1959)

Keyes, Robert W.: The Effects of Elastic Deformation on the Electrical Conductivity of Semiconductors, **11,** 149 (1960)

Inokuchi, Hiroo, and Akamatu, Hideo: Electrical Conductivity of Organic Semiconductors, **12,** 93 (1961)

Beer, Albert C.: Supplement 4–Galvanomagnetic Effects in Semiconductors (1963)

Borelius, G.: The Changes in Energy Content, Volume, and Resistivity with Temperature in Simple Solids and Liquids, **15,** 1 (1963)

Sham, L.J., and Ziman, J.M.: The Electron-Phonon Interaction, **15,** 221 (1963)

Einspruch, Norman, G.: Ultrasonic Effects in Semiconductors, **17,** 217 (1965)

Kubo, Ryogo, Miyake, Satoru J., and Hashitsume, Natsuki: Quantum Theory

of Galvanomagnetic Effect at Extremely Strong Magnetic Fields, **17**, 270 (1965)

Spector, Harold N.: Interaction of Acoustic Waves and Conduction Electrons, **19**, 291 (1966)

Conwell, Esther M.: Supplement 9–High Field Transport in Semiconductors (1967)

Keyes, Robert W.: Electronic Effects in the Elastic Properties of Semiconductors, **20**, 37 (1967)

Duke, C. B.: Supplement 10–Tunneling in Solids (1969)

Glicksman, Maurice: Plasmas in Solids, **26**, 275 (1971)

Huebener, R.P.: Thermoelectricity in Metals and Alloys, **27**, 63 (1972)

Wolf, E.L.: Nonsuperconducting Electron Tunneling Spectroscopy, **30**, 1 (1975)

Schnatterly, S.E.: Inelastic Electron Scattering Spectroscopy, **34**, 275 (1979)

Peeters, F.M., and Devreese, J.T.: Theory of Polaron Mobility, **38**, 81 (1984)

Beenakker, C.W.J., and van Houten, H.: Quantum Transport in Semiconductor Nanostructures, **44**, 1 (1991)

Weaver, J.H., and Poirier, D.M.: Solid State Properties of Fullerenes and Fullerene-Based Materials, **48**, 1 (1994)

Ferry, David K., and Grubin, Harold L.: Modeling of Quantum Transport in Semiconductor Devices, **49**, 283 (1995)

Mahan, G.D.: Good Thermoelectrics, **51**, 81 (1997)

Sorbello, Richard S.: Theory of Electromigration, **51**, 159 (1997)

12. Lattice Dynamics

de Launay, Jules: The Theory of Specific Heats and Lattice Vibrations, **2**, 219 (1956)

Klemens, P.G.: Thermal Conductivity and Lattice Vibrational Modes, **7**, 1 (1958)

Leibfried, G., and Ludwig, W.: Theory of Anharmonic Effects in Crystals, **12**, 275 (1961)

Mitra, Shashanka, S.: Vibration Spectra of Solids, **13**, 1 (1962)

Maradudin, A.A., Montroll, E.W., and Weiss, G.H.: Supplement 3—Theory of Lattice Dynamics in the Harmonic Approximation (1963)

Maradudin, A.A.: Theoretical and Experimental Aspects of the Effects of Point Defects and Disorder on the Vibrations of Crystals—1, **18**, 274 (1966)

Maradudin, A.A.: Theoretical and Experimental Aspects of the Effects of Point Defects and Disorder on the Vibrations of Crystals—2, **19**, 1 (1966)

Kwok, Philip C.K.: Green's Function Method in Lattice Dynamics, **20**, 213 (1967)

Joshi, S.K., and Rajagopal, A.K.: Lattice Dynamics of Metals, **22**, 159 (1968)

Wallace, Duane C.: Thermoelastic Theory of Stressed Crystals and Higher-Order Elastic Constants, **25**, 302 (1970)

Maradudin, A.A., Montroll, E.W., Weiss, G.H., and Ipatova, I.P.: Supplement 3, Second Edition–Theory of Lattice Dynamics in the Harmonic Approximation (1971)

Bendow, Bernard: Multiphonon Infrared Absorption in the Highly Transparent Frequency Regime of Solids, **33**, 249 (1978)

Miller, A., MacKinnon, A., and Weaire, D.: Beyond the Binaries—The Chalcopyrite and Related Semiconducting Compounds, **36**, 119 (1981)

Phillips, J.C.: Spectroscopic and Morphological Structure of Tetrahedral Oxide Glasses, **37**, 93 (1982)

Currat, R., and Janssen, T.: Excitations in Incommensurate Crystal Phases, **41**, 201 (1988)

Weaver, J.H., and Poirier, D.M.: Solid State Properties of Fullerenes and Fullerene-Based Materials, **48**, 1 (1994)

13. Phonon Transport

Mendelssohn, K., and Rosenberg, H.M.: The Thermal Conductivity of Metals at Low Temperatures, **12**, 223 (1961)

Slack, Glen A.: The Thermal Conductivity of Nonmetallic Crystals, **34**, 1 (1979)

14. Structural Defects and Atomic Transport

Seitz, Frederick, and Koehler, J.S.: Displacement of Atoms during Irradiation, **2**, 305 (1956)

Eshelby, J.D.: The Continuum Theory of Lattice Defects, **3**, 79 (1956)

Kröger, F.A., and Vink, H.J.: Relations between the Concentrations of Imperfections in Crystalline Solids, **3**, 310 (1956)

Dexter, D.L.: Theory of the Optical Properties of Imperfections in Nonmetals, **6**, 353 (1958)

Amelinckx, S., and Dekeyser, W.: The Structure and Properties of Grain Boundaries, **8**, 327 (1959)

Lazarus, David: Diffusion in Metals, **10**, 71 (1960)

Gourary, Barry S., and Adrian, Frank J.: Wave Functions for Electron-Excess Color Centers in Alkali Halide Crystals, **10**, 127 (1960)

de Wit, Roland: The Continuum Theory of Stationary Dislocations, **10**, 249 (1960)

Gilman, J.J., and Johnston, W.G.: Dislocations in Lithium Fluoride Crystals, **13**, 148 (1962)

Amelinckx, S., Supplement 6—The Direct Observation of Dislocations (1964)

Compton, W. Dale, and Rabin, Herbert: F-Aggregate Centers in Alkali Halide Crystals, **16,** 121 (1964)

Corbett, James W.: Supplement 7—Electron Radiation Damage in Semiconductors and Metals (1966)

Nichols, D.K., and van Lint, V.A.J.: Energy Loss and Range of Energetic Neutral Atoms in Solids, **18,** 1 (1966)

Maradudin, A.A.: Theoretical and Experimental Aspects of the Effects of Point Defects and Disorder on the Vibrations of Crystals–1, **18,** 274 (1966)

Maradudin, A.A.: Theoretical and Experimental Aspects of the Effects of Point Defects and Disorder on the Vibrations of Crystals–2, **19,** 1 (1966)

Markham, Jordan J.: Supplement 8—F-Centers in Alkali Halides (1966)

Alexander, H., and Haasen, P.: Dislocations and Plastic Flow in the Diamond Structure, **22,** 28 (1968)

Peterson, N.L.: Diffusion in Metals, **22,** 409 (1968)

Dederichs, P.H.: Dynamical Diffraction Theory by Optical Potential Methods, **27,** 135 (1972)

Samara, G.A.: High-Pressure Studies of Ionic Conductivity in Solids, **38,** 1 (1984)

Zunger, Alex: Electronic Structure of $3d$ Transition-Atom Impurities in Semiconductors, **39,** 275 (1986)

Nelson, David R., and Spaepen, Frans: Polytetrahedral Order in Condensed Matter, **42,** 1 (1989)

Pond, R.C., and Hirth, J.P.: Defects at Surfaces and Interfaces, **47,** 287 (1994)

Weaire, D., and McMurry, S.: Some Fundamentals of Grain Growth, **50,** 1 (1996)

Martin, Georges, and Bellon, Pascal: Driven Alloys, **50,** 189 (1996)

Sorbello, Richard S.: Theory of Electromigration, **51,** 159 (1997)

Averback, R.S., and Diaz de la Rubia, T.: Displacement Damage in Irradiated Metals and Semiconductors, **51,** 281 (1997)

15. Mechanical Properties

Eshelby, J.D.: The Continuum Theory of Lattice Defects, **3,** 79 (1956)

Rice, M.H., McQueen, R.G., and Walsh, J.M.: Compression of Solids by Strong Shock Waves, **6,** 1 (1958)

Huntington, H.B.: The Elastic Constants of Crystals, **7,** 213 (1958)

de Wit, Roland: The Continuum Theory of Stationary Dislocations, **10,** 249 (1960)

Saito, N., Okano, K., Iwayanagi, S., and Hideshima, T.: Molecular Motion in Solid State Polymers, **14,** 343 (1963)

Amelinckx, S., Supplement 6—The Direct Observation of Dislocations (1964)

Einspruch, Norman, G.: Ultrasonic Effects in Semiconductors, **17,** 217 (1965)

Doran, Donald G., and Linde, Ronald K.: Shock Effects in Solids, **19,** 229 (1966)

Keyes, Robert W.: Electronic Effects in the Elastic Properties of Semiconductors, **20,** 37 (1967)

Alexander, H., and Haasen, P.: Dislocations and Plastic Flow in the Diamond Structure, **22,** 28 (1968)

Wallace, Duane C.: Thermoelastic Theory of Stressed Crystals and Higher-Order Elastic Constants, **25,** 302 (1970)

Deuling, H.J.: Elasticity of Nematic Liquid Crystals, *in* Supplement 14—Liquid Crystals, 77 (1978)

Thomson, Robb: Physics of Fracture, **39,** 1 (1986)

Evans, A.G., and Zok, F.W.: The Physics and Mechanics of Brittle Matrix Composites, **47,** 177 (1994)

Carlsson, A.E., and Thomson, R.: Fracture Toughness of Materials: From Atomistics to Continuum Theory, **51,** 233 (1997)

16. Surfaces & Interfaces

Becker, J.A.: Study of Surfaces by Using New Tools, **7,** 379 (1958)

Montgomery, D.J.: Static Electrification of Solids, **9,** 139 (1959)

Davison, S.G., and Levine, J.D.: Surface States, **25,** 1 (1970)

Lang, Norton D.: The Density-Functional Formalism and the Electronic Structure of Metal Surfaces, **28,** 225 (1973)

Webb, M.B., and Lagally, M.G.: Elastic Scattering of Low-Energy Electrons from Surfaces, **28,** 301 (1973)

Gomer, Robert: Chemisorption on Metals, **30,** 93 (1975)

Wokaun, Alexander: Surface-Enhanced Electromagnetic Processes, **38,** 223 (1984)

Yu, Edward T., McCaldin, James O., McGill, Thomas C.: Band Offsets in Semiconductor Heterojunctions, **46,** 1 (1992)

Spaepen, Frans: Homogeneous Nucleation and the Temperature Dependence of the Crystal-Melt Interfacial Tension, **47,** 1 (1994)

Pond, R.C., and Hirth, J.P.: Defects at Surfaces and Interfaces, **47,** 287 (1994)

Osgood, Richard M. Jr., and Wang, Xiaoyi: Image States on Single-Crystal Metal Surface, **51,** 1 (1997)

17. Disordered Systems

Muto, Toshinosuke, and Takagi, Yutaka: The Theory of Order-Disorder Transitions in Alloys, **1,** 194 (1955)

Guttman, Lester: Order-Disorder Phenomena in Metals, **3,** 145 (1956)

Borelius, G.: Changes of State of Simple Solid and Liquid Metals, **6,** 65 (1958)

Saito, N., Okano, K., Iwayanagi, S., and Hideshima, T.: Molecular Motion in Solid State Polymers, **14,** 343 (1963)

Busch, G., and Güntherodt, H.-J.: Electronic Properties of Liquid Metals and Alloys, **29,** 235 (1974)

Cargill, G. S., III: Structure of Metallic Alloy Glasses, **30,** 227 (1975)

Joannopoulos, J.D., and Cohen, Marvin, L.: Theory of Short-Range Order and Disorder in Tetrahedrally Bonded Semiconductors, **31,** 71 (1976)

Ehrenreich, H., and Schwartz, L. M.: The Electronic Structure of Alloys, **31,** 149 (1976)

Ashcroft, N.W., and Stroud, D.: Theory of the Thermodynamics of Simple Liquid Metals, **33,** 1 (1978)

Wooten, F., and Weaire, D.: Modeling Tetrahedrally Bonded Random Networks by Computer, **40,** 1 (1987)

de Heer, Walt A., Knight, W.D., Chou, M.Y., and Cohen, Marvin L.: Electronic Shell Structure and Metal Clusters, **40,** 93 (1987)

McGreevy, Robert L.: Experimental Studies of the Structure and Dynamics of Molten Alkali and Alkaline Earth Halides, **40,** 247 (1987)

Nelson, David R., and Spaepen, Frans: Polytetrahedral Order in Condensed Matter, **42,** 1 (1989)

Kelton, K.F.: Crystal Nucleation in Liquids and Glasses, **45,** 75 (1991)

Yonezawa, Fumiko: Glass Transition and Relaxation of Disordered Structures, **45,** 179 (1991)

Bergman, David J.and Stroud, David: Physical Properties of Macroscopically Inhomogeneous Media, **46,** 147 (1992)

Spaepen, Frans: Homogeneous Nucleation and the Temperature Dependence of the Crystal-Melt Interfacial Tension, **47,** 1 (1994)

Evans, A.G., and Zok, F.W.: The Physics and Mechanics of Brittle Matrix Composites, **47,** 177 (1994)

Weaire, D., and McMurry, S.: Some Fundamentals of Grain Growth, **50,** 1 (1996)

Martin, Georges, and Bellon, Pascal: Driven Alloys, **50,** 189 (1996)

18. Optical Properties

Klick, Clifford C., and Schulman, James H.: Luminescence in Solids, **5,** 97 (1957)

Dexter, D.L.: Theory of the Optical Properties of Imperfections in Nonmetals, **6**, 353 (1958)

Piper, W.W., and Williams, F.E.: Electroluminescence, **6**, 95 (1958)

Givens, M. Parker: Optical Properties of Metals, **6**, 313 (1958)

McClure, Donald S.: Electronic Spectra of Molecules and Ions in Crystals. Part 1. Molecular Crystals, **8**, 1 (1959)

Newman, R., and Tyler, W.W.: Photoconductivity in Germanium, **8**, 49 (1959)

Wolf, H.C.: The Electronic Spectra of Aromatic Molecular Crystals, **9**, 1 (1959)

Scanlon, W.W.: Polar Semiconductors, **9**, 83 (1959)

McClure, Donald S.: Electronic Spectra of Molecules and Ions in Crystals. Part II. Spectra of Ions in Crystals, **9**, 399 (1959)

Mitra, Shashanka, S.: Vibration Spectra of Solids, **13**, 1 (1962)

Jørgensen, Klixbüll, Chr.: Chemical Bonding Inferred from Visible and Ultraviolet Absorption Spectra, **13**, 375 (1962)

Stern, Frank: Elementary Theory of the Optical Properties of Solids, **15**, 300 (1963)

Markham, Jordan J.: Supplement 8—F-Centers in Alkali Halides (1966)

Phillips, J.C.: The Fundamental Optical Spectra of Solids, **18**, 55 (1966)

Cardona, Manuel: Supplement 11—Optical Modulation Spectroscopy of Solids (1969)

Brown, Frederick C.: Ultraviolet Spectroscopy of Solids with the Use of Synchrotron Radiation, **29**, 1 (1974)

Nilsson, P.O.: Optical Properties of Metals and Alloys, **29**, 139 (1974)

Bendow, Bernard: Multiphonon Infrared Absorption in the Highly Transparent Frequency Regime of Solids, **33**, 249 (1978)

Phillips, J.C.: Spectroscopic and Morphological Structure of Tetrahedral Oxide Glasses, **37**, 93 (1982)

Wokaun, Alexander: Surface-Enhanced Electromagnetic Processes, **38**, 223 (1984)

Bergman, David J. and Stroud, David: Physical Properties of Macroscopically Inhomogeneous Media, **46**, 147 (1992)

Greenham, Neil C., and Friend, Richard H.: Semiconductor Device Physics of Conjugated Polymers, **49**, 1 (1995)

Hui, P.M., and Johnson, Neil F.: Photonic Band-Gap Materials, **49**,151 (1995)

Nurmikko, A.V., and Gunshor, R.L.: Physics and Device Science in II–VI Semiconductor Visible Light Emitters, **49**, 205 (1995)

Kimerling, L.C., Kolenbrander, K.D., Michel, J., Palm, J., and Giovane, L.: Light Emission from Silicon, **50**, 333 (1996)

19. Bloch Electrons in Strong External Fields

Blount, E.I.: Formalisms of Band Theory, **13,** 306 (1962)

Conwell, Esther M.: Supplement 9—High Field Transport in Semiconductors (1967)

Brown, E.: Aspects of Group Theory in Electron Dynamics, **22,** 313 (1968)

Zak, J.: The kq-Representation in the Dynamics of Electrons in Solids, **27,** 1 (1972)

20. Semiconductors

Fan, H.Y.: Valence Semiconductors, Germanium and Silicon, **1,** 283 (1955)

Welker, H., and Weiss, H.: Group III—Group V Compounds, **3,** 1 (1956)

Woodruff, Truman O.: The Orthogonalized Plane-Wave Method, **4,** 367 (1957)

Klick, Clifford C., and Schulman, James H.: Luminescence in Solids, **5,** 97 (1957)

Piper, W.W., and Williams, F.E.: Electroluminescence, **6,** 95 (1958)

Newman, R., and Tyler, W.W.: Photoconductivity in Germanium, **8,** 49 (1959)

Heiland, G., Mollwo, E., and Stöckmann, F.: Electronic Processes in Zinc Oxide, **8,** 191 (1959)

Scanlon, W.W.: Polar Semiconductors, **9,** 83 (1959)

Busch, G.A., and Kern, R.: Semiconducting Properties of Gray Tin, **11,** 1 (1960)

Keyes, Robert W.: The Effects of Elastic Deformation on the Electrical Conductivity of Semiconductors, **11,** 149 (1960)

Bube, Richard H.: Imperfection Ionization Energies in CdS-Type Materials by Photoelectronic Techniques, **11,** 223 (1960)

Lax, Benjamin, and Mavroides, John G.: Cyclotron Resonance, **11,** 261 (1960)

Ludwig, G.W., and Woodbury, H.H.: Electron Spin Resonance in Semiconductors, **13,** 223 (1962)

Beer, Albert C.: Supplement 4—Galvanomagnetic Effects in Semiconductors (1963)

Einspruch, Norman, G.: Ultrasonic Effects in Semiconductors, **17,** 217 (1965)

Corbett, James W.: Supplement 7—Electron Radiation Damage in Semiconductors and Metals (1966)

Conwell, Esther M.: Supplement 9–High Field Transport in Semiconductors (1967)

Keyes, Robert W.: Electronic Effects in the Elastic Properties of Semiconductors, **20,** 37 (1967)

Alexander, H., and Haasen, P.: Dislocations and Plastic Flow in the Diamond Structure, **22,** 28 (1968)

Dalven, Richard: Electronic Structure of PbS, PbSe, and PbTe, **28,** 179 (1973)

Joannopoulos, J.D., and Cohen, Marvin, L.: Theory of Short-Range Order and Disorder in Tetrahedrally Bonded Semiconductors, **31,** 71 (1976)

Rice, T.M.: The Electron-Hole Liquid in Semiconductors: Theoretical Aspects, **32,** 1 (1977)

Hensel, J.C., Phillips, T.G., and Thomas, G.A.: The Electron-Hole Liquid in Semiconductors: Experimental Aspects, **32,** 87 (1977)

Zunger, Alex: Electronic Structure of $3d$ Transition-Atom Impurities in Semiconductors, **39,** 275 (1986)

Isihara, A.: Electron Correlations in Two Dimensions, **42,** 271 (1989)

Carlsson, A.E.: Beyond Pair Potentials in Elemental Transition Metals and Semiconductors, **43,** 1 (1990)

Beenakker, C.W.J., and van Houten, H.: Quantum Transport in Semiconductor Nanostructures, **44,** 1 (1991)

Bastard, G., Brum, J.A., and Ferreira, R.: Electronic States in Semiconductor Heterostructures, **44,** 229 (1991)

Yu, Edward T., McCaldin, James O., and McGill, Thomas C.: Band Offsets in Semiconductor Heterojunctions, **46,** 1 (1992)

Lieber, Charles, M., and Chen, Chia-Chun: Preparation of Fullerenes and Fullerene-Based Materials, **48,** 109 (1994)

Pickett, Warren E.: Electrons and Phonons in C_{60}-Based Materials, **48,** 225 (1994)

Greenham, Neil C., and Friend, Richard H.: Semiconductor Device Physics of Conjugated Polymers, **49,** 1 (1995)

Nurmikko, A.V., and Gunshor, R.L.: Physics and Device Science in II–VI Semiconductor Visible Light Emitters, **49,** 205 (1995)

Ferry, David K., and Grubin, Harold L.: Modeling of Quantum Transport in Semiconductor Devices, **49,** 283 (1995)

Kimerling, L.C., Kolenbrander, K.D., Michel, J., Palm, J., and Giovane, L.: Light Emission from Silicon, **50,** 333 (1996)

Mahan, G.D.: Good Thermoelectrics, **51,** 81 (1997)

Sorbello, Richard S.: Theory of Electromigration, **51,** 159 (1997)

21. Magnetism

Pake, G.E.: Nuclear Magnetic Resonance, **2,** 1 (1956)

Knight, W.D.: Electron Paramagnetism and Nuclear Magnetic Resonance in Metals, **2,** 93 (1956)

Shull, C.G., and Wollan, E.O.: Applications of Neutron Diffraction to Solid State Problems, **2,** 137 (1956)

Kittel, C., and Galt, J.K.: Ferromagnetic Domain Theory, **3,** 437 (1956)

Jan, J.-P.: Galvanomagnetic and Thermomagnetic Effects in Metals, **5,** 1 (1957)

Cohen, M.H., and Reif, F.: Quadrupole Effects in Nuclear Magnetic Resonance Studies of Solids, **5,** 321 (1957)

Das, T.P., and Hahn, E.L.: Supplement 1–Nuclear Quadrupole Resonance Spectroscopy (1958)

Kahn, A.H., and Frederikse, H.P.R.: Oscillatory Behavior of Magnetic Susceptibility and Electronic Conductivity, **9,** 257 (1959)

Low, William: Supplement 2–Paramagnetic Resonance in Solids (1960)

Ludwig, G.W., and Woodbury, H.H.: Electron Spin Resonance in Semiconductors, **13,** 223 (1962)

Yafet, Y.: g Factors and Spin-Lattice Relaxation of Conduction Electrons, **14,** 1 (1963)

Anderson, Philip, W.: Theory of Magnetic Exchange Interactions: Exchange in Insulators and Semiconductors, **14,** 99 (1963)

Jarrett, H.S.: Electron Spin Resonance Spectroscopy in Molecular Solids, **14,** 215 (1963)

Hebel, L.C., Jr.: Spin Temperature and Nuclear Relaxation in Solids, **15,** 409 (1963)

Hutchings, M.T.: Point-Charge Calculations of Energy Levels of Magnetic Ions in Crystalline Electric Fields, **16,** 227 (1964)

Low, W., and Offenbacher, E.L.: Electron Spin Resonance of Magnetic Ions in Complex Oxides, Review of ESR Results in Rutile, Perovskites, Spinels, and Garnet Structures, **17,** 135 (1965)

Kubo, Ryogo, Miyake, Satoru J., and Hashitsume, Natsuki: Quantum Theory of Galvanomagnetic Effect at Extremely Strong Magnetic Fields, **17,** 270 (1965)

Nagamiya, Takeo: Helical Spin Ordering–1 Theory of Helical Spin Configurations, **20,** 305 (1967)

Cooper, Bernard R.: Magnetic Properties of Rare Earth Metals, **21,** 393 (1968)

Kittel, C.: Indirect Exchange Interactions in Metals, **22,** 1 (1968)

Kondo, J.: Theory of Dilute Magnetic Alloys, **23,** 183 (1969)

Heeger, A.J.: Localized Moments and Nonmoments in Metals: The Kondo Effect, **23,** 283 (1969)

Corciovei, A., Costache, G., and Vamanu, D.: Ferromagnetic Thin Films, **27,** 237 (1972)

MacLaughlin, Douglas E.: Magnetic Resonance in the Superconducting State, **31,** 1 (1976)

White, Robert M., and Geballe, Theodore H.: Supplement 15–Long Range Order in Solids (1979)

Fulde, Peter, Keller, Joachim, and Zwicknagl, Gertrud: Theory of Heavy Fermion Systems, **41,** 1 (1988)

Yeomans, Julia: The Theory and Application of Axial Ising Models, **41,** 151 (1988)

Bertram, H. Neal and Zhu, Jian-Gang: Fundamental Magnetization Processes in Thin-Film Recording Media, **46,** 271 (1992)

Levy, Peter M.: Giant Magnetoresistance in Magnetic Layered and Granular Materials, **47,** 367 (1994)

22. Ferroelectrics and Dielectrics

Känzig, Werner: Ferroelectrics and Antiferroelectrics, **4,** 1 (1957)

Zheludev, I.S.: Ferroelectricity and Symmetry, **26,** 429 (1971)

Zheludev, I.S.: Piezoelectricity in Textured Media, **29,** 315 (1974)

de Jeu, W.H.: The Dielectric Permittivity of Liquid Crystals, *in* Supplement 14—Liquid Crystals, 109 (1978)

Samara, G.A., and Peercy, P.S.: The Study of Soft-Mode Transitions at High Pressure, **36,** 1 (1981)

23. Superconductivity

Schafroth, M.R.: Theoretical Aspects of Superconductivity, **10,** 293 (1960)

MacLaughlin, Douglas E.: Magnetic Resonance in the Superconducting State, **31,** 1 (1976)

White, Robert M., and Geballe, Theodore H.: Supplement 15–Long Range Order in Solids (1979)

Allen, Philip B., and Mitrović, Božidar: Theory of Superconducting T_c, **37,** 1 (1982)

Tinkham, M., and Lobb, C.J.: Physical Properties of the New Superconductors, **42,** 91 (1989)

Beyers, R., and Shaw, T.M.: The Structure of $Y_1Ba_2Cu_3O_{7-\delta}$ and Its Derivatives, **41,** 135 (1989)

Hass, K.C.: Electronic Structure of Copper-Oxide Superconductors, **42,** 213 (1989)

Lieber, Charles, M. and Zhang, Zhe: Physical Properties of Metal-Doped Fullerene Superconductors, **48,** 349 (1994)

24. Alkali Halides

Gourary, Barry S., and Adrian, Frank J.: Wave Functions for Electron-Excess Color Centers in Alkali Halide Crystals, **10,** 127 (1960)

Gilman, J.J., and Johnston, W.G.: Dislocations in Lithium Fluoride Crystals, **13,** 148 (1962)

Tosi, Mario P.: Cohesion of Ionic Solids in the Born Model, **16,** 1 (1964)

Compton, W. Dale, and Rabin, Herbert: F-Aggregate Centers in Alkali Halide Crystals, **16,** 121 (1964)

Markham, Jordan J.: Supplement 8–F-Centers in Alkali Halides (1966)

25. Transition Metals & Oxides

Adler, David: Insulating and Metallic States in Transition Metal Oxides, **21,** 1 (1968)

Dimmock, J.O.: The Calculation of Electronic Energy Bands by the Augmented Plane Wave Method, **26,**103 (1971)

Sellmyer, D.J.: Electronic Structure of Metallic Compounds and Alloys: Experimental Aspects, **33,** 83 (1978)

Carlsson, A.E.: Beyond Pair Potentials in Elemental Transition Metals and Semiconductors, **43,** 1 (1990)

26. Alloys

Guinier, Andre: Heterogeneities in Solid Solutions, **9,** 293 (1959)

Nilsson, P.O.: Optical Properties of Metals and Alloys, **29,** 139 (1974)

Busch, G., and Güntherodt, H.-J.: Electronic Properties of Liquid Metals and Alloys, **29,** 235 (1974)

Ehrenreich, H., and Schwartz, L. M.: The Electronic Structure of Alloys, **31,** 149 (1976)

Sellmyer, D.J.: Electronic Structure of Metallic Compounds and Alloys: Experimental Aspects, **33,** 83 (1978)

de Fontaine, D.: Configurational Thermodynamics of Solid Solutions, **34,** 73 (1979)

Cohen, Jerome B.: The Internal Structure of Guinier-Preston Zones in Alloys, **39,** 131 (1986)

Pettifor, D.G.: A Quantum-Mechanical Critique of the Miedema Rules for Alloy Formation, **40,** 43 (1987)

de Fontaine, D.: Cluster Approach to Order-Disorder Transformations in Alloys, **47,** 33 (1994)

27. Organic Materials

Wolf, H.C.: The Electronic Spectra of Aromatic Molecular Crystals, **9,** 1 (1959)

Inokuchi, Hiroo, and Akamatu, Hideo: Electrical Conductivity of Organic Semiconductors, **12,** 93 (1961)

Jarrett, H.S.: Electron Spin Resonance Spectroscopy in Molecular Solids, **14,** 215 (1963)

Saito, N., Okano, K., Iwayanagi, S., and Hideshima, T.: Molecular Motion in Solid State Polymers, **14,** 343 (1963)

de Gennes, P.G.: Macromolecules and Liquid Crystals: Reflections on Certain Lines of Research, *in* Supplement 14–Liquid Crystals, 1 (1978)

Keller, P., and Liebert, L.: Liquid-Crystal Synthesis for Physicists, *in* Supplement 14–Liquid Crystals, 19 (1978)

Deuling, H.J.: Elasticity of Nematic Liquid Crystals, *in* Supplement 14–Liquid Crystals, 77 (1978)

de Jeu, W.H.: The Dielectric Permittivity of Liquid Crystals, *in* Supplement 14—Liquid Crystals, 109 (1978)

Dubois-Violette, E., Durand, G., Guyon, E., Manneville, P., and Pieranski, P.: Instabilities in Nematic Liquid Crystals, *in* Supplement 14–Liquid Crystals, 147 (1978)

Charvolin, Jean, and Tardieu, Annette: Lyotropic Liquid Crystals: Structures and Molecular Motions, *in* Supplement 14–Liquid Crystals, 209 (1978)

Bouligand, Y.: Liquid Crystals and Their Analogs in Biological Systems, *in* Supplement 14–Liquid Crystals, 259 (1978)

Weaver, J.H., and Poirier, D.M.: Solid State Properties of Fullerenes and Fullerene-Based Materials, **48,** 1 (1994)

Lieber, Charles, M., and Chen, Chia-Chun: Preparation of Fullerenes and Fullerene-Based Materials, **48,** 109 (1994)

Axe, J.D., Moss, S.C., and Neumann, D.A.: Structure and Dynamics of Crystalline C_{60}, **48,** 149 (1994)

Pickett, Warren E.: Electrons and Phonons in C_{60}-Based Materials, **48,** 225 (1994)

Lieber, Charles, M. and Zhang, Zhe: Physical Properties of Metal-Doped Fullerene Superconductors, **48,** 349 (1994)

Greenham, Neil C., and Friend, Richard H.: Semiconductor Device Physics of Conjugated Polymers, **49,** 1 (1995)

28. Rare Earth Metals

Cooper, Bernard R.: Magnetic Properties of Rare Earth Metals, **21,** 393 (1968)

Fulde, Peter, Keller, Joachim, and Zwicknagl, Gertrud: Theory of Heavy Fermion Systems, **41,** 1 (1988)

29. Simulations

Liu, S.H.: Fractals and Their Applications in Condensed Matter Physics, **39,** 207 (1986)

Wooten, F., and Weaire, D.: Modeling Tetrahedrally Bonded Random Networks by Computer, **40,** 1 (1987)

Yonezawa, Fumiko: Glass Transition and Relaxation of Disordered Structures, **45,** 179 (1991)

de Fontaine, D.: Cluster Approach to Order-Disorder Transformations in Alloys, **47,** 33 (1994)

Ferry, David K., and Grubin, Harold L.: Modeling of Quantum Transport in Semiconductor Devices, **49,** 283 (1995)

Carlsson, A.E., and Thomson, R.: Fracture Toughness of Materials: From Atomistics to Continuum Theory, **51,** 233 (1997)

Averback, R.S., and Diaz de la Rubia, T.: Displacement Damage in Irradiated Metals and Semiconductors, **51,** 281 (1997)

30. Data Collections

Knox, Robert S.: Bibliography of Atomic Wave Functions, **4,** 413 (1957)

Borelius, G.: Changes of State of Simple Solid and Liquid Metals, **6,** 65 (1958)

Huntington, H.B.: The Elastic Constants of Crystals, **7,** 213 (1958)

Borelius, G.: The Changes in Energy Content, Volume, and Resistivity with Temperature in Simple Solids and Liquids, **15,** 1 (1963)

Gschneidner, Karl, A. Jr.: Physical Properties and Interrelationships of Metallic and Semimetallic Elements, **16,** 275 (1964)

IV. Authors: Author Index and Article Contents

ABRIKOSOV, A.A. SUPPLEMENT 12, 1972

Introduction to the Theory of Normal Metals

A.A. ABRIKOSOV

Landau Institute for Theoretical Physics, Moscow, U.S.S.R.

Translator's Preface ix
Preface xi

I. Electrons in a Periodic Crystal Lattice
  1.1 General Properties 1
  1.2 The Tight-Binding Approximation 9
  1.3 The Nearly Free Electron Model 13

II. The Electronic Fermi Liquid
  2.1 The Quasi-Particle Concept 18
  2.2 Quasi Particles in an Isotropic Fermi Liquid . . . 20
  2.3 The Anisotropic Fermi Liquid 26
  2.4 Electronic Heat Capacity 31

III. Static Transport Coefficients
  3.1 The Electron as a Wave Packet 36
  3.2 The Kinetic Equation 39
  3.3 Electrical Conductivity 43
  3.4 Thermal Conductivity 45
  3.5 The Mean Free Path Concept 48
  3.6 Thermoelectric Effects 49
  3.7 Static Transport Coefficients of a Free-Electron Gas . . 55

IV. Scattering Processes
  4.1 Impurity Scattering 57
  4.2 Electron–Electron Scattering 60
  4.3 Scattering by Lattice Vibrations 61
  4.4 Umklapp Processes 67
  4.5 Isotope Scattering 73
  4.6 The Kondo Effect 78

V. Behavior of a Metal in a High–Frequency Electromagnetic Field
  5.1 The Normal Skin Effect 86
  5.2 The Anomalous Skin Effect—The Ineffectiveness
    Concept 88

ABR

ABRIKOSOV, A.A.

5.3 The Anomalous Skin Effect—Solution of the
Kinetic Equation 91

VI. Galvanomagnetic and Thermomagnetic Properties of Metals
6.1 The Kinetic Equation in the Presence of an Applied
Magnetic Field 104
6.2 Galvanomagnetic Effects in a Weak Magnetic Field . . 110
6.3 Galvanomagnetic Effects in a Strong Magnetic
Field—Closed Orbits 114
6.4 Galvanomagnetic Effects in a Strong Magnetic Field
and the Topology of Open Fermi Surfaces 118
6.5 Thermomagnetic Effects in Weak Magnetic Field . . 126
6.6 Thermal Conductivity and Thermomagnetic Effects in
a Stong Magnetic Field 128

VII. Cyclotron Resonance 134

VIII. Size Effects
8.1 Cutoff Cyclotron Orbits 147
8.2 Cyclotron Resonance Splashes in the High-Frequency
Field Inside a Metal 149
8.3 The Nonresonant Size Effect 152
8.4 The Nonresonant Size Effect in a Tipped Field . . . 155
8.5 The Sondheimer Effect 156
8.6 Drift Focusing of the High-Frequency Field 160
8.7 Size Effects Associated with Open Orbits 163

IX. Propagation of Electromagnetic Waves in the Presence of an Applied
Magnetic Field
9.1 Helicons in Metals with Unequal Numbers of
Electrons and Holes 165
9.2 Magnetoplasma Waves in Metals with Equal Numbers
of Electrons and Holes 168
9.3 Experimental Methods 172

X. Magnetic Susceptibility and the De Haas–Van Alphen Effect
10.1 Spin Paramagnetism 176
10.2 Level Quantization for a Free Electron in a Magnetic
Field 177
10.3 Landau Diamagnetism 180
10.4 Semiclassical Quantization of Energy Levels for an
Arbitrary Electronic System 182

ABRIKOSOV, A.A.

10.5 The De Haas–Van Alphen Effect 185
10.6 Diamagnetic Domains 194
10.7 Magnetic Breakdown. 200

XI. Quantum Effects in the High-Frequency Surface Impedance
11.1 Ordinary Quantum Oscillations 206
11.2 Resonances Associated with Skipping Orbits—
Magnetic Surface States 211

XII. Ultrasonic Attenuation in Metals
12.1 Low-Frequency Attenuation in Zero Magnetic Field . 217
12.2 High-Frequency Attenuation in Zero Magnetic Field . 223
12.3 Geometric Resonances 225
12.4 Magnetoacoustic Resonance Effects 227
12.5 Quantitative Theory of Geometric Resonances . . . 229
12.6 Quantitative Theory of Magnetoacoustic Resonance
Effects 234
12.7 Giant Oscillations in Ultrasonic Attenuation Due to
Level Quantization in a Magnetic Field 238

XIII. Fermi Liquid Effects
13.1 Interactions between Quasi Particles 245
13.2 The Landau f-Function 247
13.3 Effect of Quasi-Particle Interactions on the Magnetic
Susceptibility 251
13.4 Landau Levels and Quantum Oscillations 254
13.5 Zero Sound 256
13.6 Spin Waves 259

XIV. Methods of Band-Structure Calculations
14.1 The Orthogonalized Plane-Waves Method 268
14.2 The Pseudopotential Method 271
14.3 The Nearly Free Electron Model 274
14.4 The Dense-Matter Approximation 277

References 281

Subject Index 285

ADL–ALE

ADLER, David VOLUME 21, 1968

Insulating and Metallic States in Transition Metal Oxides

DAVID ADLER

Department of Electrical Engineering and Center for Materials Science and Engineering, Massachusetts Institute of Technology, Cambridge, Massachusetts

I. Introduction... 1
II. Theoretical Models to Explain Electrical Properties....................... 4
  1. Itinerant Models.. 4
  2. Localized Electron Models... 16
  3. Transport Properties.. 50
III. Experimental Results on Transition Metal Oxides......................... 64
  4. Semiconducting and Insulating Oxides................................. 65
  5. Metallic Oxides.. 86
  6. Oxides That Exhibit Semiconductor-to-Metal Transitions............... 91
  7. Oxides That Appear to Be Nearly Degenerate Semiconductors........... 107
IV. Conclusions... 111

ADRIAN, Frank J.: *see* GOURARY, Barry S.

AKAMATU, Hideo: *see* INOKUCHI, Hiroo

ALEXANDER, H. VOLUME 22, 1968

Dislocations and Plastic Flow in the Diamond Structure

H. ALEXANDER AND P. HAASEN

Institut für Metallphysik der Universität Göttingen, Göttingen, Germany

I. Introduction... 28
II. Dislocations in the Diamond Structure................................... 30
  1. Geometry.. 30
  2. Energetics.. 36

ALEXANDER, H.

III. Direct Observations of Dislocations in the Diamond Structure........... 38
    3. Etch Pit Studies.. 38
    4. Image Formation of Interior Dislocations......................... 45
IV. Dislocation Velocity.. 67
    5. Methods and Results.. 67
    6. Comparison with Theories of Kink Formation...................... 73
    7. The Influence of Doping.. 75
    8. Internal Friction Results.. 81
V. Dislocation Density and Multiplication............................... 83
    9. Density of Mobile Dislocations................................... 83
    10. Multiplication Laws.. 85
    11. Comparison with Experiments..................................... 87
    12. Dislocation Sources... 92
VI. Theory of the Initial Deformation...................................... 96
    13. The Effective Stress.. 96
    14. Theory of the Creep Curve....................................... 99
    15. Theory of the Yield Point....................................... 102
    16. Competitive Theories.. 104
VII. Dynamic Yield Experiments... 107
    17. Deformation Experiments... 107
    18. The Yield Region.. 109
    19. Yield Stresses.. 113
    20. Effect of Doping.. 117
    21. Yield Strains... 119
VIII. Creep Experiments.. 120
    22. The Problem of an Equation of State............................. 120
    23. Shape of the Creep Curve.. 123
    24. Stress and Temperature Dependence............................... 128
    25. Effects of Doping... 130
IX. Deformation to Higher Strains... 135
    26. The Stages of the Work-Hardening Curve.......................... 135
    27. Temperature and Strain-Rate Dependence.......................... 137
    28. Dynamic Recovery by Climb....................................... 141
    29. Fracture.. 144
X. Electrical Properties of Dislocations in the Diamond Structure....... 146
    30. Models of the Charged Dislocation............................... 146
    31. Hall Effect of Weakly Doped, Dislocated Ge...................... 149
    32. Hall Effect in Strongly Doped, Dislocated Ge and in InSb........ 152
    33. Outlook... 156

ALL

ALLEN, Philip B.   VOLUME 37, 1982

Theory of Superconducting T_c

PHILIP B. ALLEN

AND

BOŽIDAR MITROVIĆ

*Department of Physics, State Univerity of New York at Stony Brook,
Stony Brook, New York*

I. Introduction 2
   1. Preliminaries 2
   2. BCS Theory 5
II. The Normal State 11
   3. Electron–Phonon Effects 11
   4. Qualitative Discussion of Migdal's "Theorem" 22
   5. Coulomb Effects 24
   6. Impurity Effects 29
III. The Superconducting State 32
   7. Nambu Matrix Notation 32
   8. Complete Self-Energy 36
   9. Rescaling the Coulomb Interaction 38
  10. Anisotropic Equations 41
  11. Isotropic Equations 45
  12. Continuation to Real Frequencies 47
IV. Solutions for T_c 50
  13. Exact Results for Isotropic Superconductors 50
  14. Approximate T_c Equations 54
  15. Paramagnetic Impurities 59
  16. Anisotropic Superconductors 62
  17. "p-Wave" Superconductors 69
V. Complications and Speculations 73
  18. Anharmonic Effects 73
  19. Energy Dependence of Density of States 76
  20. Paramagnons 83
  21. Is There a Maximum T_c? 88
  Appendix A. The Digamma Function 89
  Appendix B. Derivation of Eq. (3.51) 91

AMELINCKX, S.   VOLUME 8, 1959

The Structure and Properties of Grain Boundaries

S. AMELINCKX AND W. DEKEYSER

Laboratory of Crystallography, Ghent, Belgium

I. Introduction.. 327
II. The Geometry of Grain Boundaries as Arrays of Dislocations........... 329
  1. Basic Models.. 329
  2. Frank's Formula... 333
  3. Applications of Frank's Formula................................. 334
    a. Boundaries Constituted by One Kind of Dislocation.............. 334
    b. Boundaries Constituted by Two Kinds of Dislocations............ 335
    c. Boundaries Constituted by Three Kinds of Dislocations........... 338
    d. Boundaries Constituted by Three Kinds of Dislocations, Generated
      by Glide.. 338
  4. Application to Specific Structures............................... 339
    a. Primitive Lattice and Related Structures: CsCl, etc.............. 339
    b. Face-Centered Cubic and Related Lattices: Thompson's Notation.. 340
    c. Possible Boundaries in the Face-Centered Lattice................ 343
    d. The Body-Centered Lattice................................. 344
  5. Singularities in Dislocation Nets............................... 349
    a. Sets of Parallel Lines: Tilt Boundaries........................ 350
    b. Square Nets... 351
    c. Hexagonal Nets.. 352
  6. Grain Boundary Junctions...................................... 354
    a. The Cellular Structure....................................... 354
    b. Stable Substructure in the Face-Centered Cubic Lattice........... 356
  7. Relation between Densities of Emergence Points at Three Intersecting
    Boundaries.. 356
III. Observations Concerning the Geometry of Grain Boundaries........... 358
  8. Surface Methods... 358
    a. Optical Reflection Microscopy................................ 358
    b. Field Emission Microscopy................................... 373
  9. Bulk Methods.. 373
    a. Sets of Parallel Lines: Tilt Boundaries........................ 378
    b. Square Nets... 379
    c. Hexagonal Patterns... 382
  10. X-Ray Methods.. 384

AME

AMELINCKX, S.

    11. Methods Suitable for Thin Crystals.............................. 385
        a. Transmission Microscopy....................................... 387
        b. Moiré Fringes.. 388
    12. Bubble Models... 391
  IV. Mechanics of Grain Boundaries as Arrays of Dislocations............... 393
    13. Interactions between Single Edge Dislocations...................... 393
    14. Stability of a Wall of Edge Dislocations.......................... 396
    15. Stress Field of an Infinite Wall of Edge Dislocations................ 397
    16. Interaction with Isolated Edge Dislocations....................... 399
    17. Interaction between Edge Dislocations Having Nonparallel Burgers
        Vectors.. 403
    18. Interaction of a Wall of Edge Dislocations and a Solute Atom........ 405
    19. Energy of a Wall of Edge Dislocations............................ 408
    20. Interaction between Single Screw Dislocations..................... 408
    21. Stress Field of an Array of Parallel Screws of the Same Sign.......... 409
    22. Stress Fields of a Crossed Grid of Screws......................... 410
    23. Interaction of Arrays of Screw Dislocations and Isolated Dislocations.. 410
    24. Energy of a Twist Boundary..................................... 414
        a. The Self-Energy... 414
        b. The Interaction Energy....................................... 415
    25. Energy of More General Grain Boundaries........................ 415
    26. Dissociation into Partials....................................... 416
    27. Energy Cusps... 417
    28. Models for Medium Angle Boundary............................. 418
    29. The Peierls Approximation Applied to Grain Boundaries............ 419
  V. Energy Measurements.. 420
    30. Principle of the Method... 420
    31. Experiments with Tricrystals.................................... 422
    32. Groove Angle Measurements.................................... 426
    33. Discussion.. 429
    34. Absolute Measurements... 430
    35. Calculation of the Energy of a High-Angle Boundary................ 432
    36. Twin Boundaries.. 434
  VI. The Generation of Sub-Boundaries................................. 435
    37. Melt-Grown Crystals.. 435
    38. Deformed Crystals. Polygonization.............................. 442
        a. Polygonization after Bending................................. 443
        b. Coarsening of Polygon Domains.............................. 445
        c. Dislocation Movement during Twist of a Specimen.............. 449
        d. "Polygonization" after Twisting.............................. 450
    39. Kinetics of Polygonization: Theory............................... 452
    40. Experimental Work on Polygonization............................ 455
        a. Geometrical Aspects... 455
        b. Kinetic Aspects... 459
  VII. Diffusion along Grain Boundaries................................. 459
 VIII. Segregation and Precipitation of Solutes at Grain Boundaries........... 468
        a. Theoretical... 468
        b. Experimental... 470
  IX. Melting at Grain Boundaries..................................... 473

AMELINCKX, S.

X. Boundary Migration. 476
  41. Introductory Remarks. 476
  42. Quantitative Formulation. 477
  43. Experimental. 479
    a. Boundary Migration during Recrystallization. 479
    b. Movement Due to Differences in State of Strain. 480
    c. Boundary Migration under an External Driving Force. 481
XI. Electrical Properties of Grain Boundaries. 491
  44. Theoretical Considerations. 491
  45. Experimental Results. Semiconductors. 492
    a. Resistance of Grain Boundaries. 492
    b. Recombination of Electrons and Holes at Small Angle Boundaries. . 495
    c. Various Considerations. 497
    d. Ionic Conductivity. 498
    e. Electroluminescence. 499

AMELINCKX, S.   SUPPLEMENT 6, 1964

The Direct Observation of Dislocations

S. AMELINCKX

Laboratory of Crystallography, Ghent, Belgium

PREFACE v

Introduction 1

Chapter I. Surface Methods

  1. Crystal Growth 3
  2. Evaporation 13
  3. Etching 15
  4. Cleavage Steps 50
  5. Slip Steps 51
  5a. Etch Spirals 53

AME

AMELINCKX, S.

Chapter II. Bulk Methods

6. Decoration Technique 55
7. X-Ray Diffraction Methods 90
8. Birefringence 105

Chapter III. Thin Film Methods

9. Electron Microscopy in Transmission: Theory 109
10. Electron Microscopy in Transmission: Applications 242
11. Direct Resolution of the Crystal Lattice 405
12. Moiré Patterns 414

Chapter IV. Field Emission and Field Ion Microscopy

13. Principle of the Method 439
14. Dislocations 440

Appendix A. Properties of Stacking Fault Fringes Taking into Account the Anomalous Absorption 441

1. Approximate Expressions for the Amplitudes of Transmitted and Scattered Beam Close to the Foil Surfaces 441
2. The Fringe Pattern Close to the Center of the Foil 447

Appendix B. Symmetry of Dislocation Images for $s = 0$, in the Absence of Absorption 463

Appendix C. 465

Author Index 471
Subject Index 479
Index of Substances 486

ANDERSON, Philip, W.   VOLUME 14, 1963

Theory of Magnetic Exchange Interactions:
Exchange in Insulators and Semiconductors

PHILIP W. ANDERSON

Bell Telephone Laboratories, Incorporated, Murray Hill, New Jersey

I. Introduction	99
II. Historical Introduction	101
III. Theories of Ferro-, Ferri-, and Antiferromagnetism	108
1. The Heisenberg Hamiltonian and Localizable Spins	108
2. Statistical Theories of Magnetism	116
IV. Spin Waves	129
V. Empirical Values of Exchange Constants in Insulators	142
VI. Theory of Superexchange	146
3. Formal Description of Magnetic Insulators: The Magnetic State	146
4. Origin of Superexchange: Kinetic Exchange	157
5. Potential Exchange and Other Exchange Effects	169
6. The Isolated Magnetic Ion: Ligand Fields and Information from Nuclear Resonance	181
7. Semiempirical Approaches and Theoretical Calculations	197
VII. Double Exchange	209
Acknowledgments	214

APPEL, J.   VOLUME 21, 1968

Polarons

J. APPEL

Gulf General Atomic, Incorporated, John J. Hopkins Laboratory for Pure and Applied Science, San Diego, California

I. Introduction	193
1. Large versus Small Polarons	195
II. Large-Polaron Eigenstates	199
2. Fröhlich Hamiltonian	199
3. Weak and Intermediate Coupling	203

APP–ASH

APPEL, J.

4. Strong Coupling 211
5. Path Integral Method 219
6. Green's Function Method 228
III. Large-Polaron Mobility 234
7. The Problem 234
8. Boltzmann Equation Approach 237
9. Density Matrix Approach 247
10. Experimental Situation 261
IV. Small-Polaron Eigenstates 276
11. Small-Polaron Bands at $T = 0$ 278
12. Small-Polaron Bands at $T \neq 0$ 292
13. Breakdown of the Small-Polaron Band Model; Localized Small Polarons 301
V. Small-Polaron Mobility 318
14. Different Transport Theories of Small Polarons 318
15. Drift Mobility from Ordinary Transport Theory 321
16. Drift Mobility from Kubo's Transport Theory 330
17. Hall Mobility 344
18. High-Frequency Mobility 355
19. Thermoelectric Power 362
20. Experimental Situation 366

ASHCROFT, N.W. VOLUME 33, 1978

Theory of the Thermodynamics of Simple Liquid Metals*

N.W. Ashcroft

Laboratory of Atomic and Solid State Physics and Materials Science Center, Clark Hall, Cornell University, Ithaca, New York

D. Stroud

Department of Physics, Ohio State University, Columbus, Ohio

I. Introduction 2
II. Formalism: Electrons and Ions 5
1. Electron–Ion Interaction: Pseudopotentials 6
2. The Hamiltonian 8
3. Reduction to a Quasi-One Component Fluid 10
4. Response Function and Dielectric Function 14
5. Effective Ion–Ion Interactions 16
6. Beyond Linear Response 21

ASHCROFT, N.W.

III. Formalism: Liquid Aspects 22
    7. Definition of Distribution Functions: Connection between Distribution
        Functions and Interparticle Interactions 23
    8. Approximations to $g(r)$ Based on Decoupling Schemes 26
    9. Approximations to $g(r)$ Based on the Direct Correlation Function 28
    10. Application to a Model System: The Hard-Sphere Liquid 31
    11. Variational Method for Calculating the Free Energy of Liquid Metals 34
IV. Applications to Pure Liquid Metals 38
    12. Pure Liquid Metals: Thermodynamic Functions 38
    13. Application to Melting in Simple Metals 44
V. Liquid Alloys 53
    14. Formalism for the Free Energy 54
    15. Phase Separation: Free Energy Approach 60
    16. Phase Separation: Fluctuation Approach 68
VI. Complex Systems 72
    Appendix: Correlation Functions 75

AVERBACK, R.S. VOLUME 51, 1997

Displacement Damage in Irradiated Metals and Semiconductors

R.S. AVERBACK

Department of Materials Science and Engineering, University of Illinois at Urbana–Champaign, Urbana, Illinois

T. DIAZ DE LA RUBIA

Chemistry and Materials Science Directorate, Lawrence Livermore National Laboratory, Livermore, California

I. Introduction 282
    1. Primary Interactions Between Energetic Particles and Solids 282
    2. Background Information 285
II. Binary Collision and Thermal Spike Models 293
    3. Binary Collision Approximation 293
    4. Thermal Spikes 302
III. Defect Production near Threshold Energies 306
    5. Molecular Dynamics Model 306
    6. Frenkel Pair Production at Low Recoil Energies 311
IV. Dynamics of Collision Cascades: Beyond BCA Models 325
    7. Collective Behavior in Cascades 325
V. Ion-Beam Mixing 335
    8. Phenomenological Models 335
    9. MD Simulations of Ion-Beam Mixing 338
    10. Experimental Results 343

AVE–AXE

AVERBACK, R.S.

VI.	Defect Production in Cascades	352
	11. Damage Function	352
VII.	Primary State of Damage	361
	12. MD Simulations of Defect Structures	361
	13. Experimental Observations of Defect Structures in Cascades	364
VIII.	Effects of Surfaces and Applied Stresses on Cascades	377
	14. MD Simulations of Ion Impacts on Surfaces	377
	15. Experimental Investigations of Ion Impacts on Surfaces	385
	16. Effect of Applied Stresses on Cascade Dynamics	390
IX.	Freely Migrating Defects	393
	17. Metals: Defect Reactions and Annealing Stages	394
	18. Freely Migrating Defects in Metals	397
	19. Freely Migrating Defects in Si: Transient-Enhanced Diffusion in Si	399
X.	Conclusions	401

AXE, J.D. VOLUME 48, 1994

Structure and Dynamics of Crystalline C_{60}

J.D. AXE

Physics Department
Brookline National Laboratory
Upton, New York

S.C. MOSS

Physics Department
University of Houston
Houston, Texas

D.A. NEUMANN

Materials Science and Engineering Laboratory
National Institute of Standards and Technology
Gaithersburg, Maryland

I.	Introduction	150
	1. Overview	150
II.	Theoretical Considerations and Some Consequences	152
	2. Statistical Mechanical Preliminaries	152
	3. Description of Molecular Orientation	154
	4. Order Parameters	157
	5. Symmetry Considerations and Landau Theory	160
	6. Mean Field Solutions	162
	7. Scattering and Orientational Disorder	166

AXE, J.D.

III.	Experimental Aspects of the Scattering from the fcc Phase	170
	8. Orientational Disorder in fcc C_{60}	170
	9. The Orientational Potential $\bar{V}(\omega)$ in the fcc Phase	179
	10. Disorder Dynamics: Reorientational Diffusion	181
	11. X-ray Diffuse Scattering	185
IV.	Experimental Aspects of the sc Phase	189
	12. Low Temperature Structure	189
	13. The Orientational Order–Disorder Transition	191
	14. Orientational Disorder in the sc Phase and the Glass Transition	198
	15. Reorientation Mechanism and the Orientational Potential in the Low Temperature Phase	202
V.	Phonons	206
	16. Experimental Aspects	206
	17. Model Calculations	215
VI.	Summary	221
	18. General Remarks	221
	Acknowledgments	222
	Glossary of Symbols	223

BASTARD, G.

VOLUME 44, 1991

Electronic States in Semiconductor Heterostructures

G. BASTARD,[†] J.A. BRUM,[*] AND R. FERREIRA[†]

†Department de Physique de l'Ecole Normale Superieure
Paris, France
*Department de Fisica
Universidade Estadual de Campinas
São Paulo, Brazil

I.	Introduction...	230
II.	The Envelope Function Approximation: A Summary	232
	1a. Bulk Band Structures	232
	1b. Building Heterostructure States................................	236
III.	Subband Edges in Quantum Wells and Superlattices.......................	250
	2. Quantum Wells..	250
	3. Superlattices	260
IV.	In-Plane Dispersion Relations	282
	4. Conduction Subbands.......................................	282
	5. Valence Subbands	292
V.	Stark Effects in Semiconductor Quantum Wells and Superlattices	302
	6. Introduction	302
	7. Electric Field Effects in Isolated Quantum Wells.....................	303
	8. Electric Field Effects in Double Quantum Wells	310
	9. Electric Field Effects in Superlattices	313

BAS–BEC

BASTARD, G.

VI. Broadening of Heterostructure Eigenstates. 337
    10. Introduction 337
    11. Carrier–Impurity and Carrier–Phonon Perturbation. . . . .
        Hamiltonians 341
    12. Electron Scattering and Transfer Times. 348
    13. Resonances in the Hole Transfer Times. 359
VII. Excitons in Semiconductor Heterostructures. 364
    14. Excitonic Effects in the Decoupled Approximation 366
    15. Coupled Excitons in Quantum Wells. 373
VIII. Quasi-One-Dimensional Systems 381
    16. Quasi-Decoupling of the Wire Eigenstates 382
    17. One-Side n-Spike- and Modulation-Doped Quantum Wires. 388
    18. Magnetoelectric Subbands in a Quasi-1D Electron Gas 393
    19. Valence Band States and Dispersion Relations. 399
    20. Quantum Wires with Lateral Defects. 407
    Acknowledgments. 415

BECKER, J.A.   VOLUME 7, 1958

Study of Surfaces by Using New Tools

J.A. BECKER

Bell Telephone Laboratories, Murray Hill, New Jersey

  I. Introduction. 379
 II. Model of a Spherical Surface of a Tungsten Single Crystal. 383
III. Experiments with the Ion Gauge. 386
    1. Adsorption and Sticking Probability. 386
    2. Stages of Desorption. 390
IV. Experiments with the Field Emission Microscope (FEM). 394
    3. Surface Migration of Tungsten Atoms. 397
    4. Effect of Adsorbed Atoms or Molecules on the Work Function. 399
    5. Adsorption of Oxygen on Single Crystal Planes. 401
    6. Adsorption and Desorption of Hydrogen on Single Crystal Planes 405
    7. The Effect of Very High Fields on Surface Migration and Desorption. . . 410
    8. Field Desorption of Hydrogen. 415
    9. Field Desorption of CH_3OH, C_2H_6, and H_2O. 419
  V. The Field Ion Microscope (FIM). 420

BEENAKKER, C. W. J.   VOLUME 44, 1991

Quantum Transport in Semiconductor Nanostructures

C.W.J. BEENAKKER and H. VAN HOUTEN

Philips Research Laboratories
Eindhoven, The Netherlands

I.	Introduction	1
	1. Preface	1
	2. Nanostructures in Si Inversion Layers	4
	3. Nanostructures in GaAs–AlGaAs Heterostructures	10
	4. Basic Properties	16
II.	Diffusive and Quasi-Ballistic Transport	26
	5. Classical Size Effects	26
	6. Weak Localization	34
	7. Conductance Fluctuations	49
	8. Aharonov–Bohm Effect	65
	9. Electron–Electron Interactions	71
	10. Quantum Size Effects	80
	11. Periodic Potential	86
III.	Ballistic Transport	98
	12. Conduction as a Transmission Problem	98
	13. Quantum Point Contacts	109
	14. Coherent Electron Focusing	125
	15. Collimation	135
	16. Junction Scattering	146
	17. Tunneling	156
IV.	Adiabatic Transport	170
	18. Edge Channels and the Quantum Hall Effect	170
	19. Selective Population and Detection of Edge Channels	181
	20. Fractional Quantum Hall Effect	202
	21. Aharonov–Bohm Effect in Strong Magnetic Fields	215
	22. Magnetically Induced Band Structure	223

BEE

BEER, Albert C. SUPPLEMENT 4, 1963

Galvanomagnetic Effects in Semiconductors

ALBERT C. BEER

Battelle Memorial Institute, Columbus, Ohio

PREFACE   v

I. Introduction   1

II. Transport Equations   3

    1. Phenomenological Equations for General Case of An-
      isotropic Solids   3
    2. Inversion of Fundamental Transport Equations   7
    3. Application to Isotropic Solids   10
    4. Physical Significance of Transport Coefficients   12
      a. One-Dimensional Case, Absence of Magnetic Field .   13
      b. Isotropic Case, Transverse Magnetic Field, Absence of
        Thermal Gradients   15

III. Expressions for Charge-Carrier Transport Coefficients. . .   18

    5. General Theory — The Boltzmann Equation   18
    6. Use of a Relaxation Time in the Boltzmann Equation   23
      a. Quadratic Energy Surfaces   24
      b. Nonquadratic Energy Surfaces   27
    7. Expressions for Transport Coefficients in Isotropic Solids   32
      a. Conductivity Coefficients   32
      b. Phenomenological Expression for Current Density
        Vector   35
    8. Transport Coefficients in Anisotropic Isothermal Solids   36
      a. Conductivity Tensors   36
      b. Resistivity Tensors   41
      c. Application of Phenomenological Relations to Various
        Crystal Systems   43
        (1) General Systems   43
        (2) Cubic Groups O_h, O, or T_d (systems of point group
          symmetry $m3m$, 432, $\bar{4}3m$)   44
        (3) System of Point-Group Symmetry $3m$ or $\bar{3}m$
          (Trigonal System C_{3v} or D_{3d})   46
        (4) Orthorhombic System (Group D_{2h})   48

IV. Experimental Determination of Transport Coefficients in
    Isothermal Media   50

    9. Measurements on Isotropic Materials   50

BEER, Albert C.

    a. Discussion of Transverse Magnetoresistance and Hall Effects 52

    b. Experimental Determinations of Hall Coefficient and Magnetoresistance 54

        (1) Effect of Transverse Current on Hall Voltage . . 56
        (2) Effect of Transverse Currents on Magnetoresistance 58
        (3) Nongeometrical Causes of Transverse Currents . 59
        (4) Temperature Gradients 66

  10. Other "Hall" and Magnetoresistance Phenomena . . . 68

    a. Various Hall Phenomena 68
        (1) Quadratic "Hall" Effect 68
        (2) Less Conventional "Hall" Phenomena 69
        (3) Concerning the Definitions of the Hall Effect . . 70

    b. Corbino Magnetoresistance, Magnetoconductivity . . 71

  11. Measurements on Anisotropic Materials 76

    a. Weak-Magnetic-Field Region 77
        (1) Diamond Cubic Symmetry 77
        (2) Other Crystal Systems 79

    b. Magnetic Fields of Arbitrary Intensity 82
        (1) General Discussion of Parameters of Interest . . 82
        (2) Hall Phenomena 84
        (3) Magnetoresistance Phenomena 86
        (4) Illustrative Examples 88

V. Transport Coefficients for Isothermal Solids Assuming a Single Parabolic Energy Band and the Relaxation Time Approximation 92

  12. General Considerations — Exact Statistics 92

    a. Expressions for Conductivity Coefficients 92
    b. Limiting Cases of Weak and of Strong Magnetic Fields 95
    c. Conductivity Mobility 98
    d. Hall Coefficient, Hall Mobility, and Magnetoresistance 99
    e. Arbitrary Magnetic Field Strengths 105

  13. Commonly Encountered Scattering Mechanisms — Exact Statistics 107
    a. Scattering by Thermal Lattice Vibrations 107
    b. Scattering by Ionized Impurities 111
    c. Mixed Scattering 116

  14. Conductivity Coefficients in the Limit of Extreme Degeneracy 120

  15. Conductivity Coefficients in the Limit of Classical Statistics 123

BEE

BEER, Albert C.

    a. Limiting Cases of Weak and of Strong Magnetic Fields   124
      (1) General Power Law Scattering 124
      (2) Acoustic Phonon Scattering 124
      (3) Ionized Impurity Scattering 125
      (4) Charge Carrier Interactions 128
      (5) Mixed Scattering 130
    b. Conductivity, Hall, and Magnetoresistance Mobilities . 139
    c. General Case of Arbitrary Magnetic Field Strengths . 142
      (1) Power-Law Scattering 143
      (2) Mixed Scattering 145

VI. Multiband Conduction 148

  16. General Considerations 148
  17. Parabolic Conduction and Valence Bands 152
    a. Two-Band Case 153
    b. Three-Band Conduction 163
    c. Extrinsic p-type Germanium and Similar Semiconductors 167

VII. Nonquadratic Energy Surfaces 175

  18. General Considerations 175
  19. Valence Bands of Germanium and Silicon 177
  20. Weak-Field Transport Coefficients 181
    a. Conductivity and Hall Effect 181
    b. Magnetoconductivity and Magnetoresistance 185
  21. Arbitrary Magnetic Field Strengths 189
    a. General Considerations 189
    b. Cubic Energy Surfaces 193
    c. Other Energy Surfaces 194
    d. Application to Germanium and Silicon 195
    e. Structure in Hall Coefficient Curves of Other Materials 212

VIII. Quadratic Energy Surfaces of the Many-Valley Type 213

  22. General Discussion 213
    a. Case of Isotropic Scattering on a Surface of Constant Energy 213
    b. Anisotropic Scattering 220
  23. Application to Specific Semiconductors 223
    a. Weak Magnetic Fields 224
      (1) Single Spheroid with Principal Axes along Coordinate Directions 224

BEER, Albert C.

  (2) Systems of Spheroids Possessing Cubic Symmetry,
    Coordinate Axes along Cube Axes   226
  (3) System of Ellipsoids Possessing Cubic Symmetry.   239
 b. Arbitrary Magnetic Field Strengths   242
  (1) Constant Mean Free Path   242
  (2) Scattering Processes Other Than $\tau \sim \varepsilon^{-1/2}$ . . .   251
 c. Noncubic Crystals   254

IX. **Special Bands and Scattering Processes**   267

 24. Nonparabolic Bands and Nonequivalent Valleys   267
  a. Nonparabolic Bands and Nonequivalent Valleys, with
   Special Consideration of III-V Compound Semiconductors   267
  b. Structurally Equivalent Valleys Rendered Nonequivalent by Perturbations   277
 25. Inapplicability of the Relaxation Time Approximation . .   278
  a. General Discussion — Use of Variational Technique .   278
  b. Application to Semiconductors   282
   (1) General Considerations   282
   (2) Criteria for Applicability of Relaxation Time Concept   283
   (3) Application of the Variational Method to Polar
    Semiconductors   289
  c. The Variational Principle in the Presence of a Magnetic
   Field   296
 26. General Treatment of Conductivity in Anisotropic Metals   301

X. **Miscellaneous Items**   308

 27. Inhomogeneities   308
  a. Random Inhomogeneities in Measurement Sample   .   308
  b. Gross Inhomogeneities in Measurement Sample . . .   312
   (1) Continuous Variations in Conductivity   312
   (2) The Step-Function Model   318
  c. Experimental Data   321
  d. Inhomogeneities in Magnetic Field   323
  e. Techniques for Reducing Distortion due to Inhomogeneities   325
  f. Non-Ohmic Effects in Inhomogeneous Specimens . .   326
 28. Quantum-Mechanical Transport Theory, Strong Magnetic
  Field Effects   328
  a. General Discussion   328
  b. Determination of Transport Coefficients   334
  c. Data on Strong-Field Magnetoresistance in Semiconductors   337
  d. Other Strong-Magnetic-Field Effects — Depopulation
   Phenomena   340
 29. Adiabatic and Isothermal Phenomena   344

BEER, Albert C.

30. More General Treatments of Scattering 354
    a. Electron-Phonon Interactions in Nonpolar Semiconductors with a Complex Band Structure 355
    b. Combined Phonon and Impurity Scattering 360
    c. Multiphonon Processes 361
    d. Other Investigations of Scattering 362
31. Strong Interactions 365
    a. Discussion of Strong Scattering 365
    b. Impurity Conduction 367
    c. Hopping Processes — "Narrow-Band" Semiconductors 372
    d. Polarons 378
32. Miscellaneous 379
    a. Non-Ohmic Effects in a Solid 379
        (1) Charge-Carrier Multiplication 380
        (2) Field-Dependent Mobilities, "Hot Electrons" . . 386
        (3) Field-Dependent Recombination Rates 390
    b. Photon Effects, PEM Phenomena 391

AUTHOR INDEX 395

SUBJECT INDEX 406

BELLON, Pascal: *see* MARTIN, Georges

BENDOW, Bernard VOLUME 33, 1978

Multiphonon Infrared Absorption in the Highly Transparent Frequency Regime of Solids

BERNARD BENDOW

Solid State Sciences Division, Rome Air Development Center, Hanscom Air Force Base, Bedford, Massachusetts

  I. Introduction 249
 II. Theory 253
    1. General Considerations 253
    2. Anharmonicity Absorption 258
    3. Nonlinear Electric Moments 271
    4. Temperature Dependence 277
    5. Sum Rules 283
 III. Experiment 285
    6. General Considerations 285
    7. Experimental Results 290
    8. Extrinsic Effects 306
 IV. Conclusions and Discussion 311

BERGMAN, David J.   VOLUME 46, 1992

Physical Properties of Macroscopically Inhomogeneous Media

DAVID J. BERGMAN

Raymond and Beverly Sackler Faculty of Exact Sciences
School of Physics and Astronomy
Tel Aviv University, Ramat Aviv
Tel Aviv, Israel

DAVID STROUD

Department of Physics
The Ohio State University
Columbus, Ohio

I.   Introduction   148
II. DC Electrical Properties—General Theory and Calculational
     Techniques   150
     1. Definitions and Basic Ideas   150
     2. Exact Results   152
     3. Simple Approximations   152
     4. Better Approximations   155
     5. Exact Bounds   158
     6. Analytical Properties and Series Expansions   161
     7. Effective-Medium Approximation   165
     8. Numerical Techniques for Continuum Composites   174
     9. Numerical Techniques for Discrete Networks   186
III. DC Electrical Properties—Applications to Specific Problems . . .   192
     10. Percolation Theory   192
     11. Dielectric Behavior Near a Percolation Threshold   198
     12. Nonuniversal Conductivity Near a Percolation Threshold . .   199
     13. Hall Effect and Magnetoresistance   201
     14. Duality in Two Dimensions   206
     15. Thermoelectricity   211
     16. Superconductivity   213
IV. Electromagnetic Properties   220
     17. Basic Equations   220
     18. Quasi-Static Approximation   222
     19. Applications   223
     20. Sum Rules   232
     21. Beyond the Quasi-Static Approximation   234
     22. Other Approximations   240

BER

BERGMAN, David J.

V. Nonlinear Properties and Flicker Noise 246
   23. Flicker Noise. 246
   24. Weak Nonlinearity 253
   25. Strong Nonlinearity. 258
   26. Nonlinear Optical Effects 267
   Acknowledgments. 269

BERTRAM H. Neal   VOLUME 46, 1992

Fundamental Magnetization Processes in Thin-Film Recording Media

H. NEAL BERTRAM

Department of Electrical and Computer Engineering and
Center for Magnetic Recording Research
University of California at San Diego, La Jolla, California

JIAN-GANG ZHU

Department of Electrical Engineering
University of Minnesota, Minneapolis, Minnesota

  I. Introduction 271
    1. Background 271
    2. Physics Overview. 276
    3. Outline. 281
 II. Microstructures and Magnetic Properties of Thin-Film Materials. 281
    4. Longitudinal Films. 282
    5. Perpendicular Films 297
III. Micromagnetic Modeling 299
    6. Review of Single- and Multiple-Particle Reversal Mechanisms 299
    7. Numerical Modeling of Thin Films 313
 IV. Reversal Processes and Domain Structures 320
    8. Longitudinal Films. 320
    9. Perpendicular Films 334
  V. Simulations of the Magnetic Recording Process 345
    10. Single Transitions. 345
    11. Transition Noise. 353
    12. Interacting Transitions 357
 VI. Self-Organized Behavior in Magnetic Systems 362
    13. Model. 363
    14. Reversal Patterns 365
    15. Correlations and Noise 369
    Acknowledgments. 371

BEYERS, R. VOLUME 41, 1989

The Structure of $Y_1Ba_2Cu_3O_{7-\delta}$ and Its Derivatives

R. BEYERS

IBM Research Division
Almaden Research Center, San Jose, California

T.M. SHAW

IBM Research Division
Thomas J. Watson Research Center, Yorktown Heights, New York

I.	Introduction	135
II.	Basic Structure	136
III.	Phase Transformations and Oxygen Arrangements	146
	1. The Orthorhombic-to-Tetragonal Phase Transition	146
	2. Ordered Oxygen Arrangements	150
	3. Phase Diagrams	155
	4. Low-Temperature Structure Changes	160
IV.	Defect Structures	162
	5. Twins	163
	6. Stacking Faults	167
	7. Grain Boundaries	170
V.	Structural Variants	173
	8. Rare Earth Substitutions	173
	9. Transition Element Substitutions	185
	10. Substitution of Other Cations	191
	11. Oxygen Substitution	192
VI.	Related Structures	194
	12. Lanthanum Copper Oxides	194
	13. Bismuth-Containing Superconductors	197
	14. Thallium-Containing Superconductors	201
VII.	Summary and Future Work	208
	Acknowledgments	211
	Notes Added in Proof	211

BLATT, Frank J. VOLUME 4, 1957

Theory of Mobility of Electrons in Solids

FRANK J. BLATT

Department of Physics and Astronomy, Michigan State University, East Lansing, Michigan

I.	Introduction	200
II.	Statistics of Free Electrons	203
	1. Extreme Degeneracy	204

BLATT, Frank J.

    2. Classical Limit.. 206
    3. Influence of Electron Interactions................................ 207
    4. Electrons in Semiconductors..................................... 208
III. Boltzmann Equation.. 214
    5. Introduction... 214
    6. Solution of the Transport Equation; Metals...................... 217
       a. Electrical Conductivity..................................... 217
       b. Thermal Conductivity....................................... 220
       c. Thermoelectric Effects..................................... 224
       d. Transverse Effects... 229
       e. Influence of Exchange and Correlation....................... 237
    7. Solution of the Transport Equation; Semiconductors.............. 238
       a. Isothermal Effects... 239
       b. Nonisothermal Effects...................................... 258
    8. Electrical Properties of Ferromagnetic Metals................... 260
       a. Resistivity.. 260
       b. Thermoelectric Properties.................................. 264
       c. Hall Effect.. 266
    9. Conduction in Thin Films....................................... 268
IV. Solution of the Boltzmann Equation for Nonspherical Energy Surfaces and
    Anisotropic Relaxation Times...................................... 271
    10. General Case.. 271
    11. The Conductivity Tensor in n-Type Germanium and Silicon........... 277
V. Scattering Mechanisms.. 287
    12. Metals... 287
       a. Lattice Scattering... 289
       b. Scattering by Static Imperfections......................... 314
       c. Solution of the Transport Equation by a Variational Method....... 328
    13. Semiconductors... 332
       a. Lattice Scattering... 332
       b. Scattering by Static Imperfections......................... 343
       c. Electron-Electron Scattering............................... 356
VI. Effects Arising from Lack of Thermal Equilibrium.................... 357
    14. "Phonon Drag" Effect.. 357
    15. The "Hot Electron" Problem.................................... 363

BLOUNT, E.I.

VOLUME 13, 1962

Formalisms of Band Theory

E.I. BLOUNT

Bell Telephone Laboratories, Murray Hill, New Jersey

I. Crystal Momentum Representation................................... 306
    1. Introduction.. 306
    2. Representation of Wave Functions and Operators.................. 308
    3. The Modified CMR.. 319

BLOUNT, E.I.

II. Wannier Representation.. 319
    4. The Wannier Hamiltonian.. 319
    5. Properties of Wannier Functions.................................... 324
    6. Problems Caused by Degeneracies.................................. 332
III. A "Mixed Representation".. 335
    7. Basic Properties.. 335
    8. Unitary Operators and Perturbation Theory......................... 337
IV. The Dirac Equation.. 343
V. Self-Consistent Field Theory... 347
VI. The $(N + 1)$-Electron Problem.. 352
    Acknowledgments.. 356
    Appendix A. The Representation of $\mathbf{x}$..................... 356
    Appendix B. Consequences of Inversion and Time Reversal............ 359
    Appendix C. Bands in Complex $\mathbf{k}$ Space.................... 362
    Appendix D. Derivation of Multiplication Rule...................... 370

BORELIUS, G.   VOLUME 6, 1958

Changes of State of Simple Solid and Liquid Metals

G. Borelius

Royal Institute of Technology, Stockholm, Sweden

I. Introduction.. 65
II. Division of Thermal Energy, Entropy, and Volume in Vibrational and Structural Parts.. 68
III. Structural Entropy and Structural Energy as Functions of the Structural Volume... 73
IV. Relationships between Structural Energy and Entropy. Melting as Order-Disorder Transformation.. 76
V. Compressibility of Cubic Metals....................................... 79
VI. Connections between the Vibrational Lattice Expansion and the Pressure Parameter... 84
VII. Change of Total Potential Energy with Change of Volume.............. 87
VIII. On the Energy of Activation for Self-Diffusion...................... 89
IX. Complications at the Melting of Metals with Structures Different from the Close-Packed Cubic One.. 91

BOR

BORELIUS, G.   VOLUME 15, 1963

The Changes in Energy Content, Volume, and Resistivity with Temperature in Simple Solids and Liquids

G. BORELIUS

Royal Institute of Technology, Stockholm, Sweden

I.	Introduction	2
II.	The Changes in Energy	4
III.	The Thermal Expansion	6
IV.	Compilation and Analysis of Results Concerning Specific Heat and Thermal Expansion	8
	1. Argon	8
	2. Krypton	11
	3. Xenon	13
	4. Mercury	14
	5. Potassium	15
	6. Sodium	18
	7. Lithium	18
	8. Tin	18
	9. Lead	20
	10. Aluminum	21
	11. Silver	24
	12. Gold	24
	13. Copper	25
	14. Nickel	27
	15. Palladium	28
	16. Platinum	28
	17. Rhodium	30
	18. Iridium	32
	19. Molybdenum	34
	20. Tungsten	35
V.	The Temperature Dependence of Resistance	36
VI.	Compilation and Analysis of Results Concerning Temperature Dependence of Resistance	38
	21. Potassium	38
	22. Sodium	39
	23. Lead	40
	24. Aluminum	40
	25. Silver	41

BORELIUS, G.

26.	Gold	42
27.	Copper	43
28.	Platinum	43
VII.	The Extended Process of Melting	46
VIII.	The Premelting	48

BOULIGAND, Y.   SUPPLEMENT 14, 1978

Liquid Crystals and Their Analogs in Biological Systems

Y. BOULIGAND

E.P.H.E. et Centre de Cytologie Experimentale, C.N.R.S., Ivry-sur-Seine, France

The aspect of mollecular pattern which seems to have been most underestimated in the consideration of biological phenomena is that found in liquid crystal.
(Needham, 1942)

I.	Introduction	259
II.	Historical Summary	262
	1. Soaps, Myelin, and Cholesterol Derivatives	262
	2. Optical Properties of Liquid Crystals and Insect Cuticle	263
	3. Geometry and Topology in Liquid Crystals and Biological Materials	263
	4. Structures of Lamellar Liquid Crystals	264
	5. Biological Membranes	265
III.	Smectics and Related Systems in Cells and Tissues	267
	6. The Electron Microscopic Investigation of Biological Materials	267
	7. Biological Bilayers	268
	8. Models	270
	9. The Liquid Character and the Morphology of Membranes	271
	10. Other Smectic Analogs	272
IV.	Nematics, Cholesterics, and Their Analogs	274
	11. Nematics and Cholesteric Inclusions in Tissues	274
	12. Stacked Rows of Parallel Arcs in Fibrous Systems	274
	13. The Example of The Crab Body Wall	274
	14. Pseudomorphoses of Cholesterics and Nematics	279
	15. Chromosomes	282
V.	Mesomorphism and Morphogenesis	289
VI.	Conclusions	293

BOY–BRO

BOYCE, J.B.: *see* HAYES, T.M.

BRILL, R. VOLUME 20, 1967

Determination of Electron Distribution in Crystals by Means of X Rays

R. BRILL

Fritz-Haber-Institut der Max-Planck-Gesellschaft, Berlin-Dahlem, Germany

I. Introduction.. 1
II. Theoretical Background.. 2
   1. Representation of Electron Distribution by Fourier Series.............. 2
   2. The Phase Problem.. 4
   3. Series Termination Effects.. 5
   4. Results on NaCl... 9
III. Determination of the Degree of Ionization by Different Methods.......... 10
   5. Atomic Scattering Curves... 10
   6. Direct Integration of Fourier Series.............................. 16
IV. Difference Fourier Synthesis.. 19
   7. Diamond.. 20
   8. Other Compounds.. 31

BROWN, E. VOLUME 22, 1968

Aspects of Group Theory in Electron Dynamics

E. BROWN

Rensselaer Polytechnic Institute, Troy, New York

I. Introduction... 313
   1. Electrons in Magnetic Fields..................................... 314
II. Charged Particle in an Electromagnetic Field.......................... 317
   2. Gauge Invariance.. 317
   3. Empty Lattice... 320
III. Group Theory... 325
   4. Vector and Ray Representations.................................. 325
   5. Vector and Ray Groups... 327
   6. Irreducible Representations..................................... 331

BROWN, E.

IV. Periodic Lattices.. 338
    7. Lattice Geometry... 338
    8. Bloch Functions and Energy Bands.................................. 340
V. Magnetic Fields... 353
    9. Magnetic Translation Group.. 353
    10. Properties of Electronic States..................................... 364
    11. Concluding Remarks... 400
    Appendix A: First Orthogonality Relations for Ray Representations........ 401
    Appendix B: Second Orthogonality Relations for Ray Representations...... 404
    Appendix C: Relation of Wannier Function to Basis Functions $\chi_{mn}{}^{q}$......... 406
    Appendix D: WKB Treatment of Interband Breakdown.................... 407

BROWN, Frederick C.   VOLUME 29, 1974

Ultraviolet Spectroscopy of Solids with the Use of Synchrotron Radiation

FREDERICK C. BROWN

Department of Physics and Materials Research Laboratory
University of Illinois, Urbana, Illinois

I. Synchrotron Radiation and the Vacuum Ultraviolet...................... 1
    1. The Ultraviolet Source Problem..................................... 3
    2. Characteristics of Synchrotron Radiation............................ 4
    3. Comparison of Synchrotron Sources................................. 9
II. Ultraviolet Response and Its Measurement............................. 12
    4. Inner Shell Energies and Hole Lifetimes............................. 12
    5. Dielectric Function and Optical Constants........................... 14
    6. Transmission and Absorption....................................... 16
    7. Reflectivity in the Vacuum Ultraviolet.............................. 23
    8. Dispersion Relations and Sum Rules................................ 27
III. Inner Shell Spectra of Recent Interest................................ 34
    9. K Absorption of the Lighter Elements............................... 34
    10. L Absorption of the Lighter Elements.............................. 43
    11. Transitions from Inner d Shells.................................... 54
    12. Rare Earth and Transition Metal Spectra........................... 63
IV. Concluding Remarks.. 70

BRU–BUB

BRUM, J.A.: *see* BASTARD G.

BUBE, Richard H. VOLUME 11, 1960

Imperfection Ionization Energies in CdS-Type Materials by Photoelectronic Techniques

RICHARD H. BUBE

RCA Laboratories, Princeton, New Jersey

I. Imperfections in Insulators... 223
  1. Introduction.. 223
  2. Imperfections and Crystal Binding................................ 224
  3. Effects on Photoconductivity..................................... 225
  4. Photoconductivity Processes...................................... 226
  5. Imperfection Sensitization of Photoconductivity.................. 229
II. Photoelectronic Techniques Applied to CdS-Type Materials.............. 230
  6. Conductivity and Hall Effect versus Temperature.................. 230
  7. Absorption... 232
  8. Excitation of Luminescence or Photoconductivity.................. 233
  9. Emission of Luminescence... 235
  10. Variation of Photoconductivity with Light Intensity............. 235
  11. Thermal Quenching of Luminescence and Photoconductivity......... 237
  12. Optical Quenching of Luminescence or Photoconductivity.......... 240
  13. Thermally Stimulated Luminescence or Photoconductivity.......... 243
  14. Space-Charge-Limited Currents................................... 247
III. Trends in the Ionization Energies in CdS-Type Materials.............. 249
  15. Summary of Measured Energies.................................... 249
  16. Approximate Correlation with Cation and Anion Constituents of the
    Material.. 249
IV. Indications of Complex Ionization Processes........................... 252
  17. Variation of Ionization Energy with Imperfection Concentration.. 252
  18. Double Levels... 253
  19. Multiple Levels... 259

BULLETT, D.W.   VOLUME 35, 1980

The Renaissance and Quantitative Development of the Tight-Binding Method

D.W. BULLETT

*Cavendish Laboratory, University of Cambridge, Cambridge, England**

I. Molecular Orbitals and the LCAO Representation 129
   1. Introduction 129
   2. Effective Hamiltonians and the Local-Density Approximation 131
   3. The LCAO Representation 133
   4. First Principles Applications of LCAO Methods 136
II. Semiempirical Methods and Model Hamiltonians 143
   5. Background to Parametrized Models 143
   6. The Extended Hückel and Neglect of Differential Overlap Methods 145
   7. LCAO as an Empirical Interpolation Scheme 151
   8. Applications of Empirical and Semiempirical Methods 154
III. Modern Quantitative Form of Tight-Binding Theory 170
   9. Historical Introduction: Wannier Functions 170
  10. The Defining Equations for More Localized Orbitals..................... 173
  11. Applications to Solids, Surfaces, and Molecules 178
IV. From Resonant Phase Shifts to Tight-Binding Parameters 192
  12. Phase Shifts and the KKR Equations 192
  13. Resonant Phase Shifts and Model Hamiltonians.......................... 195
  14. The Theory of Transition Metal Crystal Structures 202
  15. Muffin-Tin Orbitals, Atomic Spheres, and Canonical Bands 205
V. Conclusion 213

BUNDY, F.P.   VOLUME 13, 1962

Behavior of Metals at High Temperatures and Pressures

F.P. BUNDY AND H.M. STRONG

General Electric Research Laboratory, Schenectady, New York

  I. Introduction... 81
 II. Apparatus.. 82
III. Methods of Measurement... 89
IV. Pressure Calibration... 91
 V. Resistance... 94
   1. Brief Theory... 94
   2. Experimental Results... 95
VI. Compression.. 110
   3. Discussion.. 110
   4. Experimental Results.. 112

BUN–BUS

BUNDY, F.P.

VII. Phase Transformations.. 117
    5. Introduction... 117
    6. Melting Phenomena... 118
    7. $\alpha \rightarrow \gamma$ Transition in Iron.................................... 132
VIII. Effect of Pressure on Thermal emf................................... 136
    8. Introduction... 136
    9. Experimental Methods... 137
    10. Results... 140
    11. Comparison Tests... 142
IX. Role of Group VIII Metals in Diamond Synthesis..................... 143

BUSCH, G.

VOLUME 29, 1974

Electronic Properties of Liquid Metals and Alloys

G. BUSCH AND H.-J. GÜNTHERODT

Laboratorium für Festkörperphysik, ETH, Zürich, Switzerland

  I. Introduction... 235
  II. Liquid Normal Metals.. 237
    1. Introduction... 237
    2. Hall Coefficient... 238
    3. Lorenz Number.. 246
    4. Electrical Resistivity.. 250
    5. Thermal Conductivity... 262
    6. Thermopower... 264
    7. Magnetic Susceptibility... 268
    8. Density of States... 273
  III. Liquid Transition Metals.. 276
    9. Introduction.. 276
    10. Hall Coefficient.. 277
    11. Electrical Resistivity... 281
    12. Magnetic Susceptibility... 289
    13. Density of States... 301
  IV. Liquid Rare Earth Metals.. 302
    14. Introduction... 302
    15. Hall Coefficient.. 302
    16. Electrical Resistivity... 304
    17. Magnetic Susceptibility... 307
  V. Liquid Actinide Metals... 310
  VI. Conclusion.. 312

BUSCH, G.A. VOLUME 11, 1960

Semiconducting Properties of Gray Tin

G.A. Busch and R. Kern

*Laboratorium für Festkörperphysik, Eidgenössische Technische Hochschule,
Zürich, Switzerland*

I. Introduction.. 1
II. The $\alpha \leftrightarrow \beta$ Transition of Tin.. 2
   1. General Remarks.. 2
   2. The $\beta \to \alpha$ Transformation.. 3
   3. The $\alpha \to \beta$ Transformation.. 8
   4. Dependence of the Transition Temperature on Impurities and Pressure. 9
   5. Stabilization of the α Phase.. 10
   6. The α-β Element.. 10
III. Preparation of Gray Tin Specimens................................. 11
   7. Gray Tin Powder... 11
   8. Coherent Samples.. 11
   9. Single Crystals.. 12
IV. General Physical Properties.. 13
   10 Structure, Density... 13
   11. Specific Heat... 13
V. Semiconducting Properties.. 14
   12. General Remarks.. 14
   13. Experimental Results.. 17
   14. Determination of Characteristic Data......................... 28
   15. Effect of Impurities... 37
VI. Conclusions... 39

CALLAWAY, Joseph VOLUME 7, 1958

Electron Energy Bands in Solids

Joseph Callaway

Department of Physics, University of Miami, Coral Gables, Florida

I. Introduction and General Theory................................... 100
   1. The Energy Band.. 100
   2. The Many-Electron Problem.................................. 101
   3. The Hartree-Fock Equations................................. 104
   4. The Crystal Potential....................................... 106
   5. Symmetry Properties.. 108
   6. Behavior of a Band near a Degeneracy....................... 111
   7. The Density of States....................................... 113
   8. Normal Form of an Energy Band............................. 115
   9. Methods of Calculation...................................... 118

CAL

CALLAWAY, Joseph

II. The Alkali Metals... 119
   10. General Discussion.. 119
   11. Determination of the Fermi Energy......................... 121
   12. Crystal Potentials for the Alkali Metals.................... 123
   13. Core Polarization... 125
   14. Relativistic Effects....................................... 126
   15. Lithium.. 128
   16. Sodium.. 133
   17. Potassium.. 136
   18. Rubidium and Cesium..................................... 140
   19. The Knight Shift... 143
   20. Metallic Hydrogen.. 145
   21. General Survey and Conclusions........................... 146
III. Metals of Groups II and III.................................. 147
   22. General Discussion.. 147
   23. Beryllium.. 148
   24. Magnesium... 151
   25. Calcium.. 154
   26. Aluminum.. 154
   27. Solid Helium... 158
IV. Elements of Group IV and Related Semiconductors............ 158
   28. General Considerations.................................... 158
   29. Graphite... 159
   30. Diamond... 164
   31. Silicon... 168
   32. Germanium... 172
   33. Tin, Indium Antimonide, and Indium Arsenide.............. 176
   34. Gallium Arsenide... 179
V. The Transition Metals.. 180
   35. General Considerations.................................... 180
   36. Experimental Information................................. 184
   37. Titanium... 186
   38. Chromium.. 187
   39. Iron... 189
   40. Nickel... 192
   41. Copper.. 193
VI. Other Substances.. 197
   42. Silver and Gold.. 197
   43. Tungsten... 198
   44. Bismuth.. 198
   45. Selenium and Tellurium................................... 201
   46. Alkali Halides: Sodium Chloride........................... 202
   47. Potassium Chloride....................................... 205
   48. Lithium Hydride and Lithium Fluoride..................... 206
   49. Lead Sulfide... 207
   50. Zinc Sulfide... 208
   51. Barium Oxide.. 209
   52. Miscellaneous Materials................................... 210
VII. Conclusions and Prospects................................... 210

CALLAWAY, J.   VOLUME 38, 1984

Density Functional Methods:
Theory and Applications

J. CALLAWAY

Department of Physics and Astronomy, Louisiana State Unversity
Baton Rouge, Louisiana

AND

N.H. MARCH

Theoretical Chemistry Department, University of Oxford
Oxford, England

I.	Introduction	136
	1. Thomas–Fermi Method	137
	2. Dirac–Slater Local Density Treatment of Exchange	139
	3. Finite-Temperature Thomas–Fermi Theory	141
II.	The Hohenberg–Kohn Theorem and Its Extensions	143
	4. Proof of the Hohenberg–Kohn Theorem	144
	5. Spin and Relativistic Generalizations	147
	6. Finite–Temperature Theory	152
III.	Principles of Calculational Procedures	155
	7. Single-Particle Equations	156
	8. The Exchange–Correlation Functional	163
	9. Significance of the One-Electron Eigenvalues	169
	10. The Transition State	171
	11. Excited States	174
	12. The Atomic Multiplet Problem	177
	13. The Self-Interaction Correction	182
IV.	Beyond the Local Density Approximation	185
	14. Partial Summations of Gradient Series	187
	15. Length Scales and Local Density Approximation	188
	16. The One-Particle Green's Function and the Self-Energy	191
V.	Linear Response Theory	195
	17. Dielectric Function and Kohn Anomaly	195
	18. Spin Susceptibility: Paramagnets	199
	19. Spin Susceptibility: Ferromagnets	202
VI.	Some Further Applications	207
	20. Nonmagnetically Ordered Solids	207
	21. Magnetically Ordered Metals	216
VII.	Discussion	218

CAR

CARDONA, Manuel   SUPPLEMENT 11, 1969

Optical Modulation Spectroscopy of Solids

MANUEL CARDONA

Physics Department, Brown University, Providence, Rhode Island

Foreword	.. ..	v
Preface	.. ..	vii

I. **Introduction and Historical Survey** 1

II. **Optical Properties of Electrons in Solids**
    1. Introduction 9
    2. One-Electron Model 10
    3. Intraband Effects 14
    4. Critical Points and Direct Transitions 15
    5. Indirect Transitions 23
    6. Excitons 25
    7. Broadening 47
    8. Experimental Techniques 55
    9. Assignment of Optical Structure to Interband Critical Points .. 65

III. **Modulation Techniques**
    10. Fundamentals 89
    11. General Techniques 97
    12. Modulation Techniques and Band Structure Dependence on Static
    Parameters 102
    13. Kramers–Kronig Analysis 103

IV. **Wavelength Modulation**
    14. Experimental Techniques 105
    15. Line Shapes 109
    16. Results 112

V. **Temperature Modulation**
    17. General Considerations 117
    18. Experimental Techniques 123
    19. Results 125

CARDONA, Manuel

VI. **Stress Modulation**

    20. General Considerations   137

    21. Experimental Techniques: Piezoabsorption and Piezoreflectance   ..   148

    22. Results   150

VII. **Electric Field Modulation**

    23. Introduction   165

    24. Theory   166

    25. Experimental Techniques   202

    26. Experimental Results   224

VIII. **Modulation Techniques and Dependence of Band Structure on Static Parameters**

    27. Introduction: Effects of Temperature and Doping   277

    28. Electroreflectance in Binary and Pseudobinary Alloys   280

    29. Uniaxial Stress Measurements   284

    30. Modulation Techniques in the Presence of a Steady Magnetic Field   300

Appendix I.   **A Few Relationships Involving Airy Functions** ..   325

Appendix II.   **Convolution Expression for $\epsilon(\omega\,\mathscr{E})$; Experimental Absorption Edges**   329

Appendix III.   **k · p Perturbation Theory, Effective Masses, and Effective g-Factors of Semiconductors**   331

References   335

Author Index   345

Subject Index   353

CAR

CARGILL III, G.S.   VOLUME 30, 1975

Structure of Metallic Alloy Glasses

G.S. Cargill III

Department of Engineering and Applied Science, Yale University,
New Haven, Connecticut

I. Metallic Glasses... 227
   1. Introduction.. 227
   2. Metastability and Production....................................... 229
II. Description of Structure of Amorphous Solids........................... 231
   3. Three Dimensional Picture and Short Range Order.................. 231
   4. Radial Distribution Function...................................... 233
III. Summary of Experimental Techniques................................... 235
   5. X-Ray and Electron Scattering..................................... 235
   6. Neutron Scattering.. 253
   7. Electron Microscopy.. 255
   8. Crystallization and Glass Transition Behavior.................... 256
   9. Density Measurements... 257
IV. Experimental Structural Data for Metallic Glasses..................... 258
   10. Pure Metals... 258
   11. Metal–Metalloid Alloys.. 263
   12. Metal–Metal Alloys.. 281
V. Structural Models and Comparisons with Experiments................... 289
   13. Microcrystalline Models... 289
   14. Dense Random Packing of Hard Spheres........................... 295
   15. Role of Composition in Metallic Glass Formation................. 318
   16. Applications of Structural Model Results........................ 319

CARLSSON, A.E.   VOLUME 43, 1990

Beyond Pair Potentials in Elemental Transition Metals and Semiconductors

A.E. CARLSSON

Department of Physics, Washington University, St. Louis, Missouri

I.   Introduction 1
  1. Motivation and Overview 1
  2. Difficulties with Global Interatomic Potential Descriptions
     of Transition Metals and Semiconductors 11
II. Theoretical Justification and Physical Properties of Interatomic
     Potentials and Functionals 16
  3. Introduction 16
  4. Structural Energies: "Constant-Volume" Interatomic Poten-
     tials 17
  5. Broken Bonds: Pair Functional Approaches 23
  6. Cluster Functionals 42
  7. Relation Between "Constant-Volume" Potentials and Clus-
     ter Functionals 53
  8. Interatomic Potentials as Statistical Correlation Functions 57
  9. The Solid Solution Analogy 61
III. Fitted Potentials and Functionals: Implementations and Applications 64
  10. *Ab Initio* versus Fitted Descriptions 64
  11. Pair Functionals 66
  12. Angular Forces 78
IV. Concluding Comments 90

CARLSSON, A.E.   VOLUME 51, 1997

Fracture Toughness of Materials: From Atomistics to Continuum Theory

A.E. CARLSSON

Department of Physics, Washington University, St. Louis, Missouri

R. THOMSON

Emeritus National Institute of Standards and Technology, Center for Materials Science and Engineering, Gaithersburg, Maryland

I. Introduction 233
II. Levels of Description of Materials 237
  1. Atomistic Description 238

CAR–CHA

CARLSSON, A.E.

2.	Defect-Level Simulations	243
3.	Continuum Methodologies	248
III.	Calculations of Fracture Toughness	250
4.	Introduction	250
5.	Atomistic Calculations	254
6.	Defect-Level Simulations	255
7.	Continuum Calculations	261
IV.	Connections Between Different Approaches	264
8.	Atomistic to Defect-Level Connections	265
9.	Defect-Level to Continuum-Level Connections	276
V.	Key Calculations	278
10.	Defect-Level Simulations of Fracture Toughness	278
11.	Atomistic Calculation of Behavior of Dislocation Crossings	279
12.	Atomistic Calculation of the Response of Fully Three-Dimensional Crack Tips to Applied Stresses	279
13.	Development of Continuum Equations of Motion for Dislocation Densities	280

CHARVOLIN, Jean   SUPPLEMENT 14, 1978

Lyotropic Liquid Crystals: Structures and Molecular Motions

JEAN CHARVOLIN

Laboratoire de Physique des Solides, Université Paris Sud, Orsay, France

AND

ANNETTE TARDIEU

Centre de Génétique Moléculaire, C.N.R.S., Gif-sur-Yvette, France

I.	Introduction	209
II.	Structural Polymorphism	211
1.	General Properties of the Lipid Molecules	211
2.	Analysis of the X-Ray Diffraction Data	212
3.	General Description of the Polymorphism and Classification of the Structures	212
4.	Structures of the Phases	216
5.	Discussion and Comparison with Thermotropic Liquid Crystals	223
III.	Local Behavior of Lipids	225
6.	Different Approaches	225
7.	Relevant Aspects of Magnetic Resonance	227
8.	Evidence for a Dynamical Disorder	231
9.	Analysis of Lipid Motions	241
10.	Discussion	254
IV.	Final Comment	256
	Note Added in Proof	257

CHEN, Chia-Chun: *see* LIEBER, Charles M.

CHOU, M.Y.:   *see* de HEER, Walt A.

CLENDENEN, R.L.:   *see* DRICKAMER, H.G.

COHEN, Jerome B.   VOLUME 39, 1986

The Internal Structure of Guinier-Preston Zones in Alloys

JEROME B. COHEN

Department of Materials Science and Engineering
The Technological Institute
Northwestern University
Evanston, Illinois

I.	Introduction	131
	1. Metastable Stages in Alloy Decomposition	131
	2. Why Do Metastable States Exist?	133
II.	Experimental Techniques	135
	3. Imaging and Scattering Methods	135
	4. The Elastic Diffuse Scattering	137
III.	EXAFS	149
IV.	Results for Specific Alloys	150
	5. The Al–Ag System	150
	6. The Al–Zn System	156
	7. The Al–Cu System	159
V.	Discussion	186
	8. Other Systems of Interest	186
	9. Why Do Stages Occur?	188
	Appendix	196

COHEN, M.H.   VOLUME 5, 1957

Quadrupole Effects in Nuclear Magnetic Resonance Studies of Solids

M.H. COHEN AND F. REIF

Institute for the Study of Metals, Univeristy of Chicago, Chicago, Illinois

I. Introduction.. 322
   1. Background.. 322
   2. Elementary Aspects of Quadrupole Interactions.................... 324
   3. Scope of Article.. 326
II. The Nuclear Electric Quadrupole Hamiltonian....................... 326
   4. Electrostatic Interaction of the Nucleus with Its Environment....... 326
   5. The Quadrupole Hamiltonian..................................... 329
   6. Matrix Elements of the Quadrupole Hamiltonian $\mathcal{3C}_Q$.............. 332
   7. Description of Electric Field Gradients.......................... 333
III. Spectra in High Fields... 333
   8. Energy and Frequency Formulas.................................. 333
   9. Symmetric Field Gradient ($\eta = 0$)............................. 337
     a. Orientation Dependence of Frequencies...................... 337
     b. Powder Patterns... 338
     c. Rotation Patterns of Single Crystals........................ 340
   10. Asymmetric Field Gradient ($\eta \neq 0$)........................ 342
     a. Powder Patterns... 342
     b. Rotation Patterns.. 343
IV. Field Gradients.. 345
   11. Symmetry and Structural Considerations......................... 345
   12. Ionic Crystals: The Antishielding Factor........................ 346
     a. The Idealized Model.. 346
     b. The Antishielding Factor................................... 347
   13. Ionic Crystals: Local Field Gradients........................... 350
     a. Perfect Crystals... 350
     b. Imperfect Crystals... 352
   14. Ionic Crystals: Inadequacies of the Idealized Model.............. 358
   15. Remarks on Field Gradients in More General Crystal Types......... 360
   16. Imperfections in Metals.. 363
V. Static Line Broadening.. 365
   17. Internuclear Interactions in Perfect Crystals.................... 365
   18. Imperfect Cubic Crystals: Dipolar Width of the Central Component.. 366
   19. Imperfect Cubic Crystals: Statistical Theory of First-Order Broadening 368
     a. Formulation as a Statistical Problem........................ 368
     b. Line Shape for Large Concentrations........................ 370
     c. Line Shape for Small Concentrations........................ 372
     d. Intensity Loss... 375

COHEN, M.H.

20. Imperfect Cubic Crystals: Second-Order Broadening................ 376
    a. High Concentrations... 376
    b. Low Concentrations... 380
VI. Quadrupole Contribution to Nuclear Spin-Lattice Relaxation........... 383
  21. Introductory Remarks... 383
  22. General Features of Quadrupolar Relaxation.................... 385
  23. Quadrupolar Relaxation by Lattice Vibrations.................... 388
  24. Quadrupolar Relaxation by Diffusing Charges.................... 392
VII. Experimental Aspects.. 394
  25. Brief Discussion of Experimental Techniques.................... 394
VIII. Crystal Structure and Crystal Chemistry............................ 395
  26. Incidental Nuclear Information................................ 395
    a. Spin Determinations... 395
    b. Quadrupole-Moment Ratios................................. 396
    c. Quadrupole-Moment Values................................. 396
  27. Structural Information.. 396
  28. Nature of Bonding.. 402
IX. Phase Transformations... 405
  29. Phase Transformations in Ferroelectric $KNbO_3$.................... 406
  30. Phase Transformations in NH_4Br.............................. 408
X. Imperfections in Solids.. 410
  31. Dislocations.. 411
    a. Metals... 411
    b. Ionic Crystals.. 412
  32. Alloys.. 414
    a. Experimental Results.. 414
    b. Comments on Interpretation................................. 420
  33. Point Imperfections in Ionic Crystals........................... 422
    a. Small Defect Concentrations; Motion of Defects................. 422
    b. Mixed Crystals... 429
  Appendix A: The Irreducible Components of the Field Gradient Tensor.. 434
  Appendix B: Rotation Pattern Analysis by the Volkoff Method........ 435
  Appendix C: The Antishielding Factor............................ 436

COH

COHEN, Marvin L.   VOLUME 24, 1970

The Fitting of Pseudopotentials to Experimental Data and Their Subsequent Application

MARVIN L. COHEN

Department of Physics, University of California, Berkeley, California

AND

VOLKER HEINE

Cavendish Laboratory, Cambridge University, Cambridge, England

I. Why Pseudopotentials?.	38
II. Excerpts from Pseudopotential Theory.	41
1. Strong Potentials and Weaker Pseudopotentials.	41
2. Notation, Simplest Forms, and the On-Fermi-Sphere Approximation.	46
3. Screening, Variation with q, and Transferability.	51
4. Cancellation, the Empty Core, and the Delta Function Core.	54
5. Nonlocality through the Periodic Table.	57
6. Nonlocality, the Phillips–Kleinman OPW Form, and Square Wells.	59
7. Nonlocality at $G = 0$ and Effective Masses.	64
8. Augmented Plane Waves and Phase Shifts.	69
9. Spin-Orbit Coupling.	71
10. The Model Pseudopotential.	76
11. Folding Down of the Secular Equation.	78
12. Truncation of $v(q)$.	83
13. Use of $\mathbf{k} \cdot \mathbf{p}$ Perturbation Theory.	86
14. Summary Remarks, Arbitrariness of $v(q)$, and Algebraic Forms.	86
III. Fitting Procedures and the Handling of Data.	89
15. Fermi Surfaces.	90
16. Phonons.	100
17. Other Types of Data for Metals.	115
18. Optical Properties of Semiconductors and Insulators.	119
19. Results for Semiconductors and Insulators.	129
IV. Resulting Pseudopotentials for the Elements.	182
20. The Elements Discussed Individually.	182
21. Conclusions, Comparisons, and Trends.	228
V. Notes Added after the Manuscript Was Completed and in Proof.	241
Appendix A: Mathematical Notes on Nonlocal Operators.	245
Appendix B: Folding Down of the Secular Equation.	246

COHEN, Marvin L *see* de HEER, Walt A.

COHEN, Marvin L *see* JOANNOPOULOS, J.D.

COMPTON, W. Dale VOLUME 16, 1964

F-Aggregate Centers in Alkali Halide Crystals

W. DALE COMPTON

University of Illinois, Urbana, Illinois and
United States Naval Research Laboratory, Washington, D.C.

AND

HERBERT RABIN

United States Naval Research Laboratory, Washington, D.C.

I. Introduction... 121
   1. Early Investigations... 123
   2. Proposed Models... 132
II. Experimental Evidence for the Models............................. 136
   3. Optical Properties.. 136
   4. Electric Properties... 147
   5. Magnetic Properties.. 148
   6. Reaction Kinetics.. 151
III. Oscillator Strength... 161
IV. Processes of Formation and Destruction........................... 165
   7. Ionizing Radiation... 165
   8. Optical Conversion... 176
   9. Thermal Processes.. 187
V. Theory... 195
VI. Studies Relating to the First Excited States..................... 199
   10. The M Center.. 199
   11. The R and N Centers...................................... 209
VII. Studies Relating to the Higher Excited States................... 211
   12. Absorption... 211
   13. Luminescence... 220
VIII. Concluding Remarks.. 223

CON

CONWELL, Esther M.   SUPPLEMENT 9, 1967

High Field Transport in Semiconductors

ESTHER M. CONWELL

General Telephone and Electronics Laboratories, Inc., Bayside, New York
and
École Normale Supérieure, Paris, France

Preface   v
List of Symbols   xi

I. **Introduction**   1

II. **Survey of High-Field Effects**   7
    1. Effect of High Fields on the Velocity Distribution of the Carriers   . .   7
    2. Mobility in High DC Fields   15
    3. High-Frequency Conductivity in High Fields   34
    4. "Thermoelectric Effect" of Hot Carriers   45
    5. Effects Peculiar to the Many-Valley Band Structure   49
    Effects Peculiar to p-Ge   59
    Effects Due to a Magnetic Field Superimposed on a High Electric
    Field   72
    8. Negative Resistivity and the Gunn Effect   80
    9. Miscellaneous Effects   99

III. **Interaction of the Carriers with the Scattering Mechanisms**   105
    1. The Momentum Relaxation Time for Acoustic Scattering: The Simple
    Model   105
    2. The Momentum Relaxation Times for Acoustic Mode Scattering: The
    Many-Valley Model   112
    3. The Zero Point Lattice   116
    4. Rate of Energy Loss to Acoustic Modes: Simple and Many-Valley
    Models   119
    5. Disturbance of the Phonon Distribution by the Hot Carriers   127
    6. Optical Mode and Intervalley Scattering   149

IV. **Some Comparisons of Theory with Experiment**   161
    1. Deductions from Conservation of Energy   161
    2. Mobility Variations in Germanium   165
    3. The Region of Constant Drift Velocity and the Anistropy in n–Ge . .   187
    4. Polar Semiconductors   197
    5. Microwave Conductivity in High DC Fields   204

CONWELL, Esther M.

V. Formulation and Solution of the Boltzmann Equation .. 210

1. General Considerations 210
2. Collision Operator for Acoustic Modes 218
3. Collision Operator for Optical and Intervalley Scattering 221
4. Solutions for Acoustic Mode Interactions Only 225
5. Solutions for Acoustic and Optical Mode Interactions: Application to n–Ge 231
6. The Coupled Equations for Carrier and Phonon Distributions: Solution for Small Disturbances 238
7. Solution for Polar Optical and Intervalley Scattering: Application to GaAs 248
8. Treatment of Impurity Scattering and Electron-Electron Interactions 266
9. Microwave Conductivity in High DC Fields 272
10. Electromotive Force Due to an Electric Field Gradient 276

Author Index 283
Subject Index 288

COOPER, Bernard R.   VOLUME 21, 1968

Magnetic Properties of Rare Earth Metals

BERNARD R. COOPER

General Electric Research and Development Center, Schenectady, New York

I. Introduction 393

II. Equilibrium Magnetic Behavior 396
1. Equilibrium Magnetic Ordering Arrangements 396
2. Basic Model for Magnetic Properties of Rare Earth Metals 398
3. Transition from Spiral to Ferromagnetic Arrangement 406
4. Magnetic Field Effects on Equilibrium Magnetic Arrangements 413

III. Theory of Spin Waves for the Ferromagnet 417
5. Temperature Dependent Excitations for Zero Applied Field 417
6. Applied Field Effects 423
7. Magnetoelastic Effects 427

IV. Spin Waves for Periodic Moment Arrangements 437
8. Spin Waves for Cone and Planar Spiral with Zero Applied Field 437
9. Applied Field Effects on Spiral 444
10. Sinusoidal Spin Arrangement 456

V. Experimental Manifestation of Spin Waves 458
11. Neutron Inelastic Scattering Experiments 458
12. Magnetic Resonance Experiments 466
13. Macroscopic Properties 483

VI. Concluding Remarks 489

COR

CORBETT, James W.   SUPPLEMENT 7, 1966

Electron Radiation Damage in Semiconductors and Metals

JAMES W. CORBETT

General Electric Research and Development Center, Schenectady, New York

PREFACE   v

I. INTRODUCTION

1. Introduction   1
2. Qualitative Discussion of Damage Events.   2

II. BACKGROUND MATERIAL

3. Introduction   6
4. Electron-Nucleus Collision.   6

    (a) Geometry and Conservation Laws   6
    (b) Differential Cross Section for Coulomb Scattering   13

5. Cross Section for Atom Displacement   17

    (a) Step-Function Threshold Probability   17
    (b) Threshold Probability Phenomenology   22
    (c) Multiple Atom Displacement Phenomenology   25

6. Electron Multiple Scattering.   30
7. Yang Correction.   34
8. Electron Energy Loss   35
9. Annealing of Defects.   36
10. Experimental Techniques   43

    (a) Introduction   43
    (b) Carrier Lifetime in Semiconductors   44
    (c) Electrical Conductivity—Semiconductors   47
    (d) Photoconductivity   48
    (e) Electron Paramagnetic Resonance   49
    (f) Conventional Spectroscopy   55
    (g) Electrical Conductivity—Metals   55
    (h) Mechanical Relaxation Effects   56

CORBETT, James W.

| | (i) | Dilatation Effects | 57 |
| | (j) | Stored Energy | 58 |

III. ELEMENTAL SEMICONDUCTORS

11. | Silicon | 59 |
	(a)	Introduction	59
	(b)	The Vacancy [Si-G1 and Si-G2]	67
	(c)	The Divacancy [Si-G6(J) and Si-G7(C)].	71
	(d)	The Vacancy-Oxygen Pair [Si-B1(A)]	75
	(e)	The Vacancy-Phosphorus Pair [Si-G8(E)]	76
	(f)	The Vacancy-Aluminum Pair [Si-G9]	78
	(g)	Interstitial Defects	78
	(h)	The Si-G16(B) Spectrum	79
	(i)	The Si-P1(N) and Si-P3(II, III) Spectra	79
	(j)	Other EPR Spectra	83
	(k)	Electrical and Optical Measurements	83
	(l)	Length Change Measurements	88
	(m)	"Thin-Sample" Threshold Experiments	89
	(n)	"Thick-Sample" Threshold Experiments	92

12. | Germanium | 95 |
	(a)	Introduction	95
	(b)	Related Work	96
	(c)	Annealing below 80°K	101
	(d)	Annealing between 80° and 225°K	111
	(e)	Annealing above 225°K	115
	(f)	Summary of the Recovery	124
	(g)	Electronic Energy Levels	127
	(h)	Threshold Experiments	134

13. | Diamond | 144 |
	(a)	Introduction	144
	(b)	Radiation Damage Studies	149
	(c)	Theory of Radiation-Induced Defects in Diamond	155
	(d)	EPR Measurements on Irradiated Diamonds	157

14. | Graphite | 162 |
| | (a) | Introduction | 162 |
| | (b) | Radiation Damage Studies | 165 |

15. | Tellurium | 173 |

COR

CORBETT, James W.

IV. COMPOUND SEMICONDUCTORS

16. Introduction 176
17. Indium Antimonide 181
18. Indium Arsenide 186
19. Gallium Arsenide 187
20. Gallium Antimonide 191
21. Cadmium Sulfide 192
22. Zinc Selenide 196
23. Other Compound Semiconductors 197

V. PURE METALS

24. Introduction 199
25. Copper 206

    (a) Introduction 206
    (b) Stage I ($T < 60°K$) 206
    (c) Stage II ($60°K < T < 220°K$) 219
    (d) Stages III and IV ($220°K < T < 450°K$) 226
    (e) Critique of the Recovery Models 232
    (f) Threshold Experiments 240

26. Gold 243

    (a) Introduction 243
    (b) Vacancy Information 243
    (c) Radiation Damage and Plastic Deformation Experiments ... 246
    (d) Critique of the Recovery Models 254
    (e) Threshold Experiments 259

27. Silver 260

    (a) Vacancy Information 260
    (b) Radiation Damage and Plastic Deformation Experiments ... 261
    (c) Recovery Models 264
    (d) Threshold Experiments 265

28. Aluminum 266

    (a) Vacancy Information 266
    (b) Radiation Damage and Plastic Deformation Experiments: Recovery Models 267
    (c) Threshold Experiments 271

29. Nickel 272

    (a) Stages I and II 272
    (b) Stages III and IV 274

CORBETT, James W.

VI. ALLOYS

CORCIOVEI, A.   VOLUME 27, 1972

Ferromagnetic Thin Films

A. Corciovei, G. Costache, and D. Vamanu

Institute for Atomic Physics, Bucharest, Romania

I.	Introduction	237
II.	Magnetic Energy	238
	1. Macroscopic Description	238
	2. Experimental Data on the Magnetic Energy of Single Crystals	251
	3. Microscopic Description	256
III.	Domains and Ripple	266
	4. Magnetization in Thin Single Domain Specimens	267
	5. Domain Structure	275
	6. Magnetization Ripple	283
IV.	Saturation Magnetization	291
	7. Molecular Field and High Temperature Expansions	292
	8. Spin Waves	313
	9. Green's Function Method	341
V.	Concluding Remarks	349

COSTACHE, G.;   *see* CORCIOVEI, A.

CURRAT, R.   VOLUME 41, 1988

Excitations in Incommensurate Crystal Phases

R. CURRAT

*Institut Laue-Langevin
Centre de Tri 156X,
Grenoble, France*

T. JANSSEN

*Institute for Theoretical Physics,
University of Nijmegen,
Toernooiveld,
Nijmegen, The Netherlands*

I.		Introduction	202
	1.	Incommensurate Crystal Phases as Quasiperiodic Structures	202
	2.	Modulated Crystals	204
	3.	Real Incommensurate Solids	205
	4.	Systems and Mechanisms	207
II.		Landau Theory of Modulated Systems	211
	5.	Normal–Incommensurate Phase Transitions	211
	6.	Discommensurations	213
	7.	Excitations in the Incommensurate Phase	217
	8.	Debye–Waller Factor	221
	9.	Light Scattering from Incommensurate Phases	223
III.		Supersymmetry and Higher-Dimensional Space Groups	225
	10.	Symmetry of Quasiperiodic Structures	225
	11.	Excitations	228
	12.	Inelastic Scattering	231
	13.	Transformation Properties and Selection Rules	234
IV.		Microscopic Models	236
	14.	Frenkel–Kontorova Model	236
	15.	DIFFFOUR Model	238
	16.	Excitations in Incommensurate Phases: Simple Models	242
	17.	Dynamics of the DIFFFOUR Model	249
	18.	Dynamics of the Frenkel–Kontorova Model	256
	19.	Optical Properties	257
V.		Long-Wavelength Excitations in Composite Systems	260
	20.	Structure and Excitations	260
VI.		Experimental Results	264
	21.	Neutron Scattering Results on Modulated Crystals	264
	22.	Optical Studies	287
	23.	Results from Other Techniques	297
VII.		Concluding Remarks	301

DAL–DAS

DALVEN, Richard VOLUME 28, 1973

Electronic Structure of PbS, PbSe, and PbTe

RICHARD DALVEN

Department of Physics, University of California, Berkeley, California

I. Introduction 179
   1. Aim and Scope 179
   2. The Crystal Lattice 181
   3. The Brillouin Zone and Symmetry 182
   4. Crystal Bonding and Ionicity 184
   5. Phonon Spectra 186
   6. Dielectric Properties 188
II. Band Structure 191
   7. Calculated Band Structures 191
   8. Ordering of Energy Levels at the L Point 197
   9. The Minimum Energy Gap E_0 198
   10. Temperature and Pressure Dependence of E_0 203
   11. Structure of the Conduction and Valence Bands near L ... 207
   12. Bands at Other Energies 216
   13. Other Aspects of Electronic Structure 221
III. Summary 223

DAS, T.P. SUPPLEMENT 1, 1958

Nuclear Quadrupole Resonance

T.P. DAS

Saha Institute of Nuclear Physics, Calcutta, India

E.L. HAHN

Department of Physics, University of California, Berkeley, California

Preface. v

Introduction 1

I. Theory 3

DAS, T.P.

1. Frequencies and Intensities of Pure Quadrupole Spectra 3
   a. Hamiltonian 3
   b. Pure Quadrupole Spectra for Axially Symmetric Field Gradients 5
   c Zeeman Splitting of the Quadrupole Spectra. Case of Axially
     Symmetric Field Gradient 7
   d. Pure Quadrupole Spectra with Nonaxial Field Gradients. . . . 12
   e. Zeeman Splitting for the Nonaxial Case 19
2. Static Splitting and Broadening of Quadrupole Spectra. 26
   a. Splitting and Broadening of Spectra Due to Magnetic Interactions 26
   b. Electrical Sources of Broadening 39
3. Effects of Internal Motions in Molecular Solids on Nuclear Quadrupole Resonance 39
   a. Effects of Torsional Motions of Molecules 40
   b. Effects of Hindered Internal Rotations on Pure Quadrupole
      Resonance. 64
4. Theory of Transient Experiments in Quadrupole Resonance. . . . 70
   a. Pure Quadrupole Spin Echo and Free Induction Signals 71
   b. Spin Echo and Free Induction Signals in the Presence of a Weak
      Magnetic Field. 77
   c. Relative Advantages of Steady-State and Transient Experiments 80

II. Instrumentation 83

5. Apparatus 83
   a. General Considerations 83
   b. Regenerative Continuous Wave Detectors 84
   c. Super-Regenerative Method of Detection 88
   d. Circuits for High-Frequency Bromine and Iodine Resonances. . 91
   e. Circuitry for Nuclear Quadrupole Echoes and Free Induction . 93

III. Applications of Interest to the Solid State. 97

6. Information on the Constitution of the Crystalline Unit Cell . . . 98
   a. Number of Chemically Inequivalent Lattice Sites in the Unit Cell 98
   b. Interpretation of Zeeman Splittings and Their Use in Determin-
      ing Orientations of Molecules in the Unit Cell 102
   c. Study of Phase Transitions by Quadrupole Resonance. 113
7. Interpretation of Nuclear Quadrupole Coupling Data in Terms of
   Electron Distribution in Free Molecules 119
   a. Townes and Dailey Theory for Interpretation of Nuclear Quadru-
      pole Coupling Constants in Free Molecules. 120
   b. Relation between Asymmetry Parameter and Double Bond
      Characters of Halogen Bonds in Planar Molecules. 154
8. Intermolecular Binding in the Solid State. 164
   a. Relative Importance of Direct and Indirect Effects of Neighbor-
      ing Molecules on the Nuclear Quadrupole Interaction 164
   b. Differences in Ionic Characters of Chemical Bonds in the Gaseous
      and Solid States 166
   c. Bond Switching. 170
   d. Interpretation of Quadrupole Resonance Spectra in Solid Halo-
      gens in Terms of the Intermolecular Binding. 173

DAS, T.P.

    9. Quadrupole Resonance Studies of Nuclei Other Than Halogens . . 179
        a. Compounds of Group I Elements of the Periodic Table 180
        b. Compounds of Second Group Elements 183
        c. Compounds of Third Group Elements. 185
        d. Compounds of Fifth Group Elements 192
   10. Broadening of Quadrupole Resonance Lines by Impurities 205
        a. General Considerations 205
        b. Quadrupole Resonance in Irradiated Specimens. 207
        c. Quadrupole Resonance in Solid Solutions 208

Author Index 211

Subject Index 215

DAVISON, S.G.

VOLUME 25, 1970

Surface States

S.G. DAVISON

Quantum Theory Group, Departments of Applied Mathematics and Physics
University of Waterloo, Ontario, Canada

J.D. LEVINE

RCA Laboratories, David Sarnoff Research Center, Princeton, New Jersey

   I. Introduction. 2
  II. Brief Historical Review. 3
 III. Crystal Orbital Methods. 5
    1. Direct Approach. 5
    2. Resolvent Technique. 18
  IV. Crystal Potential Methods. 31
    3. Kronig–Penney Model. 32
    4. Mathieu Problem. 54
   V. Three-Dimensional Calculations. 75
    5. Direct Extension of One-Dimensional Picture. 76
    6. Analytical Continuation. 77
    7. Complications at the Boundary Region. 78
    8. Density of States. 79
  VI. Semiclassical Methods. 80
    9. Dangling Bond Approach. 81
   10. Madelung Potential Method. 84
 VII. Comparison between Theory and Experiment. 88
   11. Basic Phenomena. 90
   12. Surface State Distributions. 94
   13. Experimental Methods. 102
   14. Surface Preparation. 108
   15. Characterization by LEED. 110
   16. Line Defects. 112
   17. Data and Interpretation. 113
VIII. Concluding Remarks. 148
   List of Abbreviations. 149

DED

Dynamical Diffraction Theory by Optical Potential Methods

P.H. DEDERICHS

Institut für Festkörperforschung der Kernforschungsanlage Jülich
Jülich, Germany

I. Introduction 136
II. Diffraction of Electrons... 138
　　1. Bloch Waves.. 138
　　2. Complex $\mathbf{k}$, E and Symmetries of $E_\nu(\mathbf{k})$............................. 141
　　3. Band Structure for Complex k (One Dimension) 144
　　4. Band Structure for Complex $\mathbf{k}$ (Three Dimensions and $V \neq V^\dagger$)........ 149
　　5. The Two-Beam Case... 156
　　6. Some Multiple-Beam Cases.. 161
　　7. Orthogonality of Bloch Functions on the Dispersion Surface............ 165
III. Diffraction of Low-Energy Electrons, Neutrons, and X-Rays.............. 166
　　8. Multiple Scattering with Single-Scattering Matrices.................. 166
　　9. Scattering by Muffin-Tin Potentials.............................. 170
　　10. Diffraction of Neutrons... 173
　　11. Diffraction of X-Rays 179
IV. The Optical Potential.. 184
　　12. The Coherent Wave 184
　　13. Properties of the Optical Potential............................... 188
　　14. Approximations for the Optical Potential.......................... 193
V. Application to Electron Diffraction................................... 197
　　15. The Averaged Potential 197
　　16. Electronic Excitations... 200
　　17. Phonon Excitations 204
VI. Optical Potential Method for X-Ray Diffraction........................ 210
　　18. The Photoelectric Effect... 210
　　19. Influence of Thermal Motion..................................... 216
VII. Influence of Statistically Distributed Defects......................... 221
　　20. Isolated Point Defects... 222
　　21. Effects of Defect Clustering 228

DEF

DE FONTAINE, D. VOLUME 34, 1979

Configurational Thermodynamics of Solid Solutions

D. DE FONTAINE

School of Engineering and Applied Science, University of California, Los Angeles, California

I.	Introduction	74
II.	State of Order	75
	1. Site Occupation	75
	2. Short-Range and Long-Range Order	77
	3. Concentration Waves	80
	4. Replacive Scattering	82
	5. Discussion of "State of Order"	86
III.	Internal Energy	88
	6. Pair Potentials	90
	7. Singularities of $V(h)$	99
	8. Symmetry of $V(h)$	104
	9. Continuum Gradient Expansion	112
	10. Discussion of Internal Energy	114
IV.	Elastic Interactions	116
	11. Microscopic Elastic Theory	117
	12. Continuum Elastic Theory	124
	13. Particle Interactions	130
	14. Displacive Scattering	138
	15. Discussion of Elastic Interactions	143
V.	Free Energy Models	146
	16. Configurational Free Energy	146
	17. Generalized Bragg–Williams Model	148
	18. Taylor and Fourier Expansions	154
	19. Landau–Lifshitz Theory	157
	20. Cluster Variation Method	164
	21. Discussion of Free Energy Models	175
VI.	Stable and Unstable States	179
	22. Ordered Ground States	180
	23. Coherent Phase Diagrams	194
	24. Stability Analysis	210

Cluster Approach to Order-Disorder Transformations in Alloys

D. DE FONTAINE

Department of Materials Science and Mineral Engineering, University of California, Berkeley, California and Materials Sciences Division, Lawrence Berkeley Laboratory, Berkeley, California

I.	Introduction	33
II.	Statement of Problem	38
	1. Separation of Degrees of Freedom	39
	2. Variational Principle	41
III.	Cluster Expansions	43
	3. Configuration Variables	43
	4. Orthonormal Expansion	48
	5. Effective Cluster Interactions	53
	6. Discussion of Cluster Expansions	62
IV.	Ground State Analysis	64
	7. Ground State Formalism	65
	8. Ground State Examples	71
V.	Configurational Free Energy	79
	9. CVM Formalism	80
	10. Prototype Phase Diagrams	89
	11. Concentration Waves	99
VI.	ECI Calculations	108
	12. Direct Configurational Averaging	109
	13. Methods Based on the CPA	114
	14. Structure Inversion Method	118
VII.	Selected Applications	130
	15. Applications of the DCA	130
	16. Applications of the SIM	145
	17. Survey of Applications	163
VIII.	Conclusions	174
	Acknowledgments	176

DE GENNES, P.G.   SUPPLEMENT 14, 1978

Macromolecules and Liquid Crystals: Reflections on Certain Lines of Research

P.G. DE GENNES

Collège de France, Paris, France

I.   Introduction 1
II. Sheet Structures of AB Copolymers 5
     1. The Dense Side (A) 6
     2. The Semidilute Side (B) 6
     3. The Balance between Bulk and Surface Energies 8
III. Dilute Micelles of AB Copolymers 9
IV. One Ideal Chain in a Layered System 11
     4. Weak Stretching in Adverse Regions 11
     5. Strong Stretching in Adverse Regions 13
V.   A Real Chain in a Layered System 14
VI. Conclusions 15
     6. Scaling Laws for Block Copolymer/Solvent Systems 15
     7. Future Experiments on Chains Trapped in a Smectic Matrix 16
     8. Chains Confined in Rod Structures 17

DE HEER, Walt A.   VOLUME 40, 1987

Electronic Shell Structure and Metal Clusters

WALT A. DE HEER AND W.D. KNIGHT

Department of Physics, University of California, Berkeley, California

M.Y. CHOU AND MARVIN L. COHEN

Department of Physics, University of California, and Materials and Molecular Research Division, Lawrence Berkeley Laboratory, Berkeley, California

I. Development of Cluster Physics   94
   1. Historical Perspective   94
   2. Recent Developments   95
   3. Types of Clusters   97
II. Basic Electronic Shell Theory of Metal Clusters   98
   4. Introduction   98
   5. Quantum Chemistry and Solid-State Physics   100
   6. Spherical Jellium Model and Shell Structure   103
   7. Results for Some Simple Metal Clusters   105
   8. The Spheroidal Model   114
   9. Limits of Extended Shell Theory   115

DE HEER, Walt A.

III. Production and Detection of Metal Clusters		117
10. Introduction		117
11. Supersonic Nozzle Sources		117
12. Adiabatic Expansion in Free Jets		118
13. Cluster Formation in Pure Vapor Jets		119
14. Seeded Nozzle Sources		121
15. Gas Aggregation Sources		124
16. Ovenless Sources		125
17. Detectors		126
18. Overview of a Cluster Beam Apparatus		128
19. Conclusions		129
IV. Cluster Abundance Spectra and Shell Structure		130
20. Introduction		130
21. Pure Alkali Metals		130
22. Noble-Metal Ions		135
23. Mixed Clusters		140
24. Polyvalent Metals		142
25. Conclusions		143
V. Static Dipole Polarizability		144
26. Metallic Surface Properties		144
27. Theories of Polarizability		144
28. Theory for Open-Shell Clusters		148
29. Effects of Spheroidal Distortions		148
30. Experimental Polarizabilities		149
31. Comparison of Theory and Experiment		151
32. Conclusions		151
VI. Ionization Potential		152
33. Introduction		152
34. Ionization Threshold Measurements		153
35. Theoretical Framework		156
36. Ionization Potential and Cluster Size		157
37. Behavior of the Ionization Potential at Shell Closings		160
38. Effects of Spheroidal Distortions		162
39. Conclusions		164
VIII. Prospects for Metal-Cluster Physics		164
40. Introduction		164
41. Cluster Production, Structure, and Stability		165
42. Cluster Properties		167
43. Cluster Spectroscopy		168
44. New Materials		169
45. Conclusions		170
Appendix A. Applications of Nilsson Theory for Clusters		170
46. The Nilsson Diagram		170
47. Energy Scales		172
48. Predictions of the Nilsson Model		173
Appendix B. Fine Structure in Cluster Abundance Spectra		175
49. Equilibrium Distributions		175
50. Evaporation and Fragmentation		178

DEJ–DEK

DE JEU, W.H.   SUPPLEMENT 14, 1978

The Dielectric Permittivity of Liquid Crystals

W.H. DE JEU

Phillips Research Laboratories, Eindhoven, The Netherlands

I.   Introduction 109
II. The Static Permittivity of Nematic Liquid Crystals 113
     1. Nematics with Nonpolar Molecules.................................... 113
     2. Extension of the Onsager Theory to Nematics 118
III. Permittivity and Molecular Structure 122
     3. *p*-Azoxyanisole... 122
     4. Variation of the Terminal Groups in *p,p'*-di-substituted Azobenzenes 124
     5. Variation of the Bridging Group for Fixed Terminal Substituents 126
IV. The Influence of Smectic Order on the Static Permittivity 130
     6. Extension of the Kirkwood–Fröhlich Theory to Liquid Crystals 131
     7. Dipole Correlation in Nematics and Smectics 132
V.   Dynamic Behavior of the Permittivity 136
     8. Extension of Debye's Theory of Dielectric Relaxation to Nematics 137
     9. Experimental Results and Discussion 140

DEKEYSER, W.:   *see* AMELINCKX, S.

DEKKER, A.J.   VOLUME 6, 1958

Secondary Electron Emission

A.J. DEKKER

Department of Electrical Engineering, University of Minnesota, Minneapolis, Minnesota

I.   Introduction.. 251
II. Experimental Data on the Total Yield................................ 254
     1. The Yield of Metals... 254
     2. The Yield of Elemental Semiconductors......................... 257
     3. The Yield of Insulators....................................... 258
     4. Reduced Yield Curves.. 262
III. Semiempirical Theory of Secondary Emission......................... 263
     5. Assumptions of the Elementary Theory.......................... 263
     6. Theory Based on Power-Law for Primary Energy Losses........... 264
         a. Reduced Yield Curves...................................... 264
         b. Calculation of Characteristic Parameters.................. 268
     7. Theory Including Scattering of Primaries...................... 269

DEKKER, A.J.

IV. Backscattering, Reflection, and Characteristic Energy Losses of Primaries.. 272
  8. Backscattering for Primary Energies >200 eV...................... 273
    a. The Backscattering Coefficient.................................. 273
    b. The Energy Distribution of Backscattered Electrons.............. 274
  9. Sternglass' Theory of Secondary Emission......................... 276
  10. Reflection of Primary Electrons below 30 eV...................... 279
    a. Metals.. 279
    b. Insulators.. 281
  11. Characteristic Energy Losses of Primaries........................ 283
 V. Theories of the Secondary Emission of Metals........................ 286
  12. The Free Electron Theory of Baroody............................. 287
  13. Quantum Theory of the Production of Secondaries in Metals.......... 289
  14. The Cascade Theory for Metals................................... 292
  15. Secondary Emission as a Surface Effect........................... 295
VI. Miscellaneous Properties of Secondary Emission....................... 297
  16. The Secondary Yield as Function of the Angle of Incidence........... 297
  17. Temperature Dependence of the Yield of Insulators.................. 300
  18. The Influence of Lattice Defects on the Yield...................... 303
  19. The Angular Distribution of Secondary Electrons................... 305
  20. The Time Lag of Secondary Emission.............................. 305
  21. Field-Dependent Secondary Emission.............................. 306
  22. Fine Structure of the Velocity Distribution of True Secondaries........ 311

DE LAUNAY, Jules   VOLUME 2, 1956

The Theory of Specific Heats and Lattice Vibrations

JULES DE LAUNAY

U.S. Naval Research Laboratory, Washington, D.C.

  I. The Dulong-Petit Law... 220
 II. Einstein's Theory of Atomic Heat................................. 221
III. Debye's Theory of Atomic Heat.................................... 225
IV. Analysis of Experimental Data with Debye's Theory.................. 235
 V. Simple One-Dimensional Born-von Kármán Lattice.................... 237
VI. One-Dimensional Born-von Kármán Lattice with a Basis.............. 243
VII. Central and Noncentral Forces.................................... 248
  1. Central or Radial Forces....................................... 248
  2. Angular Forces... 249
  3. Noncentral Forces 251
VIII. Methods for Computing $g(\nu)$ and Related Physical Quantities........... 251
  4. The Secular Equation.. 251
  5. Brillouin Zones... 252
  6. Equations Defining $g(\nu)$....................................... 253
  7. Numerical Calculation of $g(\nu)$................................. 255
  8. The Computation of C_V and Θ................................ 257
IX. Two-Dimensional Born-von Kármán Lattice.......................... 257
  9. Formulation of the Equations of Motion......................... 258
  10. Long Wave Approximation and the Velocity of Sound.............. 260
  11. The Spectrum... 260

DE LAUNAY, Jules

X. Velocity of Sound and Crystal Elasticity.......................... 263
    12. Hooke's Law.. 263
    13. The Dynamical Equations of Sound........................... 265
    14. Velocity of Sound and the Elastic Constants.................... 266
XI. Body-Centered and Face-Centered Cubic Metallic Elements........... 269
    15. Central Force Models: the Secular Equations.................... 269
    16. Central Force Models: the Force Constants..................... 272
    17. Central Force Models: Numerical Calculations................... 274
    18. The Electron Gas Modification: Acoustic Waves.................. 276
    19. The Electron Gas Modification: Thermal Waves.................. 284
    20. Numerical Tables and Comparison with Experiment............... 285
XII. The Diamond Lattice.. 290
    21. The Diamond Structure.. 291
    22. The Secular Equation of the Model with Three Force Constants..... 293
    23. The Secular Equation of the Model with Two Force Constants...... 294
    24. The Force Constants of the Three-Constant Model................ 295
    25. Comparison of the Calculated Optical Frequency and the Results of Experiments on Infrared Absorption............................ 297
    26. Calculations of Specific Heat and Comparison with Experiment...... 298

DEULING, H.J.   SUPPLEMENT 14, 1978

Elasticity of Nematic Liquid Crystals

H.J. DEULING

Gesamthochschule Kassel, Kassel, Germany

  I. Introduction 77
 II. Freedericksz Transition in Nematics 79
    1. Distortion by Magnetic Fields....................................... 79
    2. Distortion by Electric Fields 85
    3. Freedericksz Transition in Crossed Electric and Magnetic Fields 90
III. Flexo-electric Effects 92
    4. Distortion by a Magnetic Field 92
    5. Distortion by an Electric Field 93
    6. Flexo-electric Instability 95
IV. Observation of Distortion and Determination of Material Constants 96
    7. Thermal Conductivity Method 96
    8. Electrical Conductivity Method 98
    9. Capacitance Method 98
    10. Optical Methods... 100
  V. Dynamics of the Freedericksz Transition 103

DEVREESE, J.T.:   *see* PEETERS, F.M.

DE WIT, Roland   VOLUME 10, 1960

The Continuum Theory of Stationary Dislocations

ROLAND DE WIT

University of Illinois, Urbana, Illinois

I. Introduction.. 249
   1. General Comments... 249
   2. Survey of Topics.. 251
II. Review of Elasticity Theory... 252
   3. Basic Equations of the General Theory........................... 252
   4. Green's Tensor Function.. 254
III. Early Developments of Dislocation Theory........................... 257
   5. Definition of the Dislocation.................................... 257
   6. Burgers' Formula.. 260
   7. Peach and Koehler's Formula.................................... 263
   8. The Expression for the Energy of a Dislocation.................... 265
   9. The Compatibility Equations.................................... 266
IV. Kröner's Theory.. 267
   10. Introduction... 267
   11. Review of Magnetostatics....................................... 268
   12. Kröner's Theory of Internal Stresses............................ 270
   13. Analogy between Magnetostatics and Kröner's Theory............... 273
   14. Application to Dislocations..................................... 274
   15. The Expressions for the Energy................................. 277
   16. Concluding Remarks about Self-Energy........................... 280
V. Energy Expressions for Some Simple Geometries....................... 282
   17. Straight Dislocations... 282
   18. Circular Dislocations... 286
   19. The Cylindrical Dislocation..................................... 288
   20. The Helical Dislocation... 290
Appendix. Expressions for M_{33} in Cartesian and Cylindrical Coordinates....... 291

DEXTER, D.L. VOLUME 6, 1958

Theory of the Optical Properties of Imperfections in Nonmetals

D.L. DEXTER

University of Rochester, Rochester, New York

I. Interaction of Electromagnetic Radiation with Matter.................. 355
  1. Basic Formalism.. 355
  2. Specialization to Centers in a Dielectric Medium................... 360
II. Common Simplifications and Approximations......................... 362
  3. The One-Electron Approximation.................................. 362
  4. The Static Lattice... 362
    a. Effective Mass.. 362
    b. Index of Refraction... 362
    c. Effective Field Ratio.. 363
  5. The Vibrating Lattice... 364
    a. The Adiabatic Approximation................................ 364
    b. The Condon Approximation................................... 367
  6. Oscillator Strengths.. 370
III. Specialization for Idealized Centers............................... 374
  7. The Linear Approximation....................................... 375
  8. The Tight-Binding Approximation................................ 383
    a. The Franck-Condon Principle................................ 387
    b. The Calculation of Configuration Coordinate Diagrams........... 391
    c. Experimental Determination of C.C. Diagrams.................. 394
IV. Special Topics... 395
  9. Anisotropy Effects.. 395
  10. Effects of Dislocations... 400
  11. Colloidal Particles.. 402
  12. Reflection and Excitation Spectra............................... 403
  13. Nonradiative Transitions.. 405
    a. Configuration Coordinate Description.......................... 406
    b. The Diffuse Center... 410

DIAZ DE LA RUBIA, T.: *see* AVERBACK, R.S.

DIMMOCK, J.O.   VOLUME 26, 1971

The Calculation of Electronic Energy Bands by the Augmented Plane Wave Method

J.O. DIMMOCK

Lincoln Laboratory, Massachusetts Institute of Technology, Lexington, Massachusetts

I.	Introduction	104
II.	The Energy Band Method as an Independent-Electron Model for the Electronic Properties of Solids	107
	1. The Many-Body Problem	107
	2. Electron–Phonon Interaction	108
	3. The Hartree–Fock Approximation	109
	4. Averaging of the Exchange Term	112
III.	The APW Method	118
	5. Nonrelativistic Calculations	121
	6. Relativistic Calculations	126
IV.	Comparison between the APW Method and Other Methods of Calculating Energy Bands in Solids	135
	7. Relationship between the APW and KKR Methods	137
	8. The APW and KKR Methods and Their Relationship to the Methods of Pseudopotentials and the Method of Scattered Waves	145
V.	Survey of APW Calculations and Results	149
	9. Face-Centered-Cubic Transition and Noble Metals (Cu, Ag, Au, Ni, Pd, Pt, Rh, Ir, and Solid Ar)	150
	10. Body-Centered-Cubic Transition Metals (Fe, Cr, Mo, W, V, Nb, and Ta)	182
	11. Hexagonal-Close-Packed Transition Metals (Sc, Ti, Zr, Hf, Re, Ru, Os, and Co)	194
	12. Nearly Free-Electron Metals	200
	13. Rare Earth Metals	223
	14. Semiconductors, Insulators, Ionic Compounds, and Alloys	245
VI.	Recent and Projected Future Directions	252
	15. Modifications in the Method	252
	16. Modifications in the Treatment of Exchange	253
	17. Modifications in the Muffin-Tin Potential	254
	18. Areas of Probable Endeavor	255
Appendix I.	Bibliography of Papers Pertaining to the APW Method of Calculating Energy Bands in Solids	260
Appendix II.	Bibliography of Papers Pertaining to the KKR or Green's Function Method of Calculating Energy Bands in Solids	269
	Note Added in Proof	273

DOR

DORAN, Donald G. VOLUME 19, 1966

Shock Effects in Solids

DONALD G. DORAN AND RONALD K. LINDE

Poulter Laboratories, Stanford Research Institute, Menlo Park, California

I. Introduction and Summary... 230
II. Shock Waves in Solids.. 233
    1. Definition and Production of Shock Waves.......................... 233
    2. Conservation Equations and Equations of State..................... 235
    3. Experimental Techniques.. 236
    4. Continuum Models of Shock Compression............................ 238
    5. Lattice Models of Shock Compression.............................. 241
III. Transient Observation of Material under Shock Compression.............. 243
    6. The P-V Hugoniot and its Implications............................ 244
    7. Phase Transitions.. 245
    8. Stress Relaxation.. 248
    9. Electrical Conductivity.. 250
    10. Electrical Polarization and Depolarization...................... 253
    11. Other Electrical Effects.. 256
    12. Magnetic Effects.. 257
    13. Luminescence and Optical Absorption............................. 258
    14. Mechanical Effects.. 261
IV. Terminal Observation of Shocked Material............................. 262
    15. Recovery Problems and Residual Effects.......................... 262
    16. Phase Transitions... 265
    17. Microstructure.. 266
    18. Mechanical Properties... 269
    19. Stored Energy... 272
    20. X-Ray Diffraction... 274
    21. Lattice Sublimation... 275
    22. Color Centers... 275
    23. Thermoluminescence.. 278
    24. Exoelectron Emission (Kramer Effect)............................ 280
    25. Electrical Conductivity... 281
    26. Magnetic Properties... 282
    27. Other Effects... 284
    28. Summary Listing of Residual Effects Observed or Predicted....... 286
V. Comparison with Static Work.. 286
VI. Relating Laboratory Experiments to Specimens Recovered in the Field.... 288

DRICKAMER, H.G.   VOLUME 17, 1965

The Effect of High Pressure on the Electronic Structure of Solids

H.G. DRICKAMER

Department of Chemistry and Chemical Engineering and Materials Research Laboratory,
University of Illinois, Urbana, Illinois

I. Introduction.. 1
II. Nonmetals.. 2
  1. Impurity Centers in Ionic Crystals................................. 2
  2. Crystal Field Effects...................................... 19
  3. Band Structure and the Approach to the Metallic State................. 34
  4. Organic Crystals... 57
III. Metals.. 89
  5. Electronic and Metal–Nonmetal Transitions......................... 91
  6. The Electronic Structure of Hexagonal Close-Packed Metals............ 112
  7. The Electronic Structure of Iron at High Pressure..................... 127

DRICKAMER, H.G.   VOLUME 19, 1966

X-Ray Diffraction Studies of the Lattice Parameters of Solids under Very High Pressure

H.G. DRICKAMER, R.W. LYNCH, R.L. CLENDENEN, AND E.A. PEREZ-ALBUERNE

Department of Chemistry and Chemical Engineering and Materials Research Laboratory,
University of Illinois, Urbana, Illinois

I. Introduction... 135
II. Ionic and Ionic-Covalent Crystals............................... 140
  1. Alkali Halides.. 143
  2. Miscellaneous Halides.................................. 149
  3. Cubic Oxides and Sulfides............................... 151
  4. Cubic Carbides....................................... 160
  5. Tetragonal Oxides..................................... 160
  6. The Corundum Structure................................ 162
III. Molecular and Covalent Crystals.............................. 167
  7. Diamond, Graphite, and Boron Nitride...................... 167
  8. Iodine, SnI_4, and $p\text{-}C_6H_4I_2$.......................... 180
IV. Metals and Alloys.. 191
  9. Hexagonal Close-Packed Metals and Alloys.................. 192
  10. Indium and Its Tetragonal Alloys......................... 204
  11. Cubic Alloys... 208
  Appendix A... 210
  Appendix B... 228

DUB

DUBOIS-VIOLETTE, E.   SUPPLEMENT 14, 1978

Instabilities in Nematic Liquid Crystals

E. Dubois-Violette, G. Durand, E. Guyon, P. Manneville, and P. Pieranski

Laboratoire de Physique des Solides, Université Paris-Sud, Orsay, France

I.	Introduction	147
II.	Linear Hydrodynamics of Isotropic and Nematic Liquids	148
	1. Intermediate Geometries	152
III.	Bénard–Rayleigh Thermal Convection in Isotropic Fluids	153
IV.	Thermal Convection in Nematics	156
	2. Planar Case	156
	3. Homeotropic Convection	163
V.	Hydrodynamic Instabilities in Nematics	167
	4. Experiments	167
	5. Theoretical Approach	173
	6. Other Flows	183
VI.	Electrohydrodynamic (EHD) Instabilities in Nematics	188
	7. Introduction	188
	8. The Helfrich and DGP Model	189
	9. Time-Dependent Analysis	193
	10. Generalization to Different Materials and Geometries	197
	11. Pretransitional Phenomena	201
	12. Discussion and Conclusion	203
	Appendix: Analog Simulation of Linear Instabilities	205

122

DUKE, C.B. SUPPLEMENT 10, 1969

Tunneling in Solids

C.B. DUKE

General Electric Research and Development Center, Schenectady, New York
now at
Department of Physics, University of Illinois, Urbana, Illinois

Foreword v

Preface vii

I. Introduction 1

II. Historical Commentary 4

III. Models of the Barrier Hamiltonian
1. Survey of Types of Models 14
2. Models Emphasizing Electron-Electron Interactions 15
3. Models Emphasizing Band-Structure Effects 22
4. The Phenomenological Point of View 27

IV. The Independent-Electron Theory of Tunneling
5. The Metal-Vacuum Contact 30
6. Internal Field Emission 36
   a. Types of Internal Field Emission 36
   b. One-Electron Schrödinger Equations 38
   c. Qualitative Features of Models for Zener Emission 39
   d. Survey of the Literature on Zener Emission 45
7. Heterojunction Tunneling 49
   a. Evaluation of the Current and Conductance 49
   b. Influence of Band Structure on Tunneling Characteristics 54
   c. Metal-Insulator-Metal Junctions 57
   d. Metal-Insulator-Semiconductor (Semimetal) Junctions 65
   e. Metal-Semiconductor Contacts 71
   f. Magnetic Fields and Composite Junctions 77
8. The p-n Tunnel Diode 80
   a. Specular Direct Tunneling 80
   b. Magnetic-Field Effects: Theory and Experiment 85

V. Comparison of the Consequences of the Independent-Electron Model with Experiments
9. Metal-Insulator-Metal Junctions 90
10. Metal-Insulator-Semiconductor and Metal-Insulator-Semimetal Junctions 97
11. Metal-Semiconductor Contacts 102

DUK

DUKE, C.B.

12. Semiconductor Tunnel Heterodiodes 110
13. p-n Tunnel Diodes: I—Elastic Tunneling 112
   a. General Features of Tunnel-Diode Current-Voltage Characteristics 112
   b. Direct Diodes 117
   c. Indirect Diodes 120

VI. Beyond the Independent-Electron Model

14. p-n Tunnel Diodes: II—Inelastic Tunneling 122
   a. Phonon-Assisted Tunneling in Indirect Diodes 122
   b. Phonon-Assisted Tunneling in Direct Diodes 125
   c. Photon-Assisted Tunneling in p-n Diodes 127
15. Survey of Experimental Evidence for the Influence of Electron
   Interactions on Tunneling 135
16. Single-Particle Tunneling into Superconducting Electrodes 167
   a. Survey of Experimental Measurements 167
   b. Elastic Tunneling: Special Topics 170
17. The Josephson Effect 193
   a. Properties of Weakly Coupled Superconductors 193
   b. Macroscopic Model of an Ideal Tunnel Junction 196
   c. Experimental Confirmation of the Theoretical Predictions 201

VII. Many-Body Theory of Electron Tunneling

18. Definition of the Transfer-Hamiltonian Model 207
   a. Conceptual Foundations 207
   b. Coherent Elastic Tunneling through an Average Potential Barrier .. 215
   c. Fluctuation-Induced Elastic Tunneling 220
   d. Occupation-Number Representation of the Transfer Hamiltonian .. 222
19. Linear-Response Formalism for the Tunneling Current 230
   a. The Current as a Functional of Temperature Green's Functions .. 230
   b. Diagrammatic Expansion of the Single-Particle Current 237
   c. Diagrammatic Expansion of the Josephson Supercurrent 245
20. Many-Body Interactions in the Electrodes: I—Normal Electrodes .. 250
   a. The Single-Electrode Case 250
   b. Electron-Phonon Interactions in Homopolar Electrodes 255
   c. Electron Interactions in Polar-Semiconductor Electrodes 263
21. Many-Body Interactions in the Electrodes: II—Superconducting
   Electrodes 269
   a. The Pairing Model 269
   b. Single-Particle Tunneling 273
   c. Supercurrent Flow (Condensed-Pair Tunneling) 276
22. Inelastic Tunneling 279
   a. Qualitative Features 279
   b. Phonon Emission in Indirect Semiconductors 282
   c. Phonon Emission in Direct Semiconductors 286
   d. Inelastic Impurity Excitation in the Barrier 290
   e. Resonant Tunneling via Paramagnetic Impurities 294

References 308

Author Index 327
Subject Index 341

DURAND, G.: *see* DUBOIS-VIOLETTE, E.

ECHENIQUE, P.M. VOLUME 43, 1990

Dynamic Screening of Ions in Condensed Matter

P.M. ECHENIQUE

Departamento de Física de Materiales
Facultad de Ciencias Químicas
Universidad del País Vasco/Euskal Herriko Unibertsitatea
San Sebastián, Spain

F. FLORES

Departamento de Física del Estado Sólido
Universidad Autónoma de Madrid
Cantoblanco, Madrid, Spain

R.H. RITCHIE

Health and Safety Research Division
Oak Ridge National Laboratory
Oak Ridge, Tennessee

I.	Introduction	230
	1. Background	231
	2. Historical Sketch	231
	3. Approach	236
II.	Quantal Self-Energy of a Projectile in an Electron Gas	236
	4. Semiclassical Self-Energy for a Charge Moving in a Uniform Electron Gas	236
	5. Quantal Self-Energy of a Moving Charge	239
	6. Quantal Self-Energy for an External Electron in an Electron Gas	242
	7. Bound Ion–Electron Composite Interacting with an Electron Gas	246
	8. Interactions in which the Internal State of the Composite Is Unchanged	248
III.	Linear Theory of the Interaction of a Probe with Condensed Matter	250
	9. Wake Concept	250
	10. Plasmon Dispersion	252
	11. Single-Particle Effects, Plasmon Pole Approximation	253
	12. Wake in the RPA	256
	13. Damping Effects	259

ECH

ECHENIQUE, P.M.

126

EHRENREICH, H.   VOLUME 31, 1976

The Electronic Structure of Alloys

H. EHRENREICH

Division of Engineering and Applied Physics, Harvard University, Cambridge, Massachusetts

AND

L.M. SCHWARTZ

Physics Department, Brandeis University, Waltham, Massachusetts

I. Introduction... 150
  1. Early Models... 151
  2. Overview... 156
II. Formulation of the Problem.. 159
  3. Configuration Averages.. 159
  4. Single Particle Green's Functions............................... 162
  5. Component Densities of States................................... 167
III. Multiple Scattering Theory....................................... 169
  6. The Basic Equations... 169
  7. The Single Site Approximations.................................. 173
IV. The Single Band Alloy... 176
  8. The Perfect Crystal... 176
  9. Definition of the Single Band Alloy............................. 182
  10. The Weak Scattering Limit...................................... 183
  11. The Dilute Alloy... 185
  12. The Localization Theorem....................................... 189
  13. Moments.. 190
  14. Comparison of the CPA and ATA.................................. 196
  15. Equilibrium Properties in the CPA.............................. 204
  16. The Steeple Model.. 209
V. Formal Extensions of the Theory................................... 211
  17. Diagrammatic Methods... 211
  18. Random Transfer Integrals...................................... 217
VI. The Alloy Potential.. 224
  19. Atomic Potentials and Charge Transfer.......................... 224
VII. Realistic Model State Densities................................. 227
  20. Definition of the Model.. 227
  21. Applications to CuNi and AgPd Alloys........................... 231
VIII. The s–d Model.. 238
  22. Definition of the Model.. 238
  23. The CPA.. 240
  24. Application to AuAg Alloys..................................... 243
IX. The Muffin Tin Model... 253
  25. Perspective.. 253
  26. The Angular Momentum Representation............................ 255
  27. The Multiple Scattering Formalism.............................. 258
  28. The Single Site Approximations................................. 264
  29. Applications to CuZn and CuNi Alloys........................... 271
  Appendix: Reduction of the Multiple Scattering Equations to Their
  Energy Shell Components.. 283

EIN–ESH

EINSPRUCH, Norman G.   VOLUME 17, 1965

Ultrasonic Effects in Semiconductors

NORMAN G. EINSPRUCH

Physics Research Laboratory, Texas Instruments Incorporated, Dallas, Texas

I. Introduction.. 217
II. Summary of Classical Analysis.. 218
    1. Hooke's Law, Elastic Constants, Wave Equation..................... 218
    2. Scattering.. 224
    3. Diffraction... 225
    4. Definitions... 225
III. Measurement Techniques.. 226
    5. Megacycle Techniques—Pulse Echo............................... 226
    6. Gigacycle Techniques—Microwave Ultrasonics...................... 228
    7. Samples and Bonds.. 229
IV. Semiconductors... 230
    8. Electronic Attenuation... 230
    9. Some Generalizations on the Elastic Moduli....................... 235
    10. Electronic Contribution to the Shear Moduli...................... 237
    11. Acoustoelectric Effect.. 243
    12. Piezoelectric Interaction.. 247
    13. Radiation Damage.. 264
    14. Phonon–Phonon Interactions.. 265
    General References.. 268

ESHELBY, J.D.   VOLUME 3, 1956

The Continuum Theory of Lattice Defects

J.D. ESHELBY

University of Birmingham, Birmingham, England

I. Introduction.. 79
    1. Relation between Lattice Defects and Continuum Theory............ 79
    2. Basic Ideas and Survey of Topics.................................... 80
II. Formal Theory... 84
    3. Elements of Elasticity... 84
    4. Specification of Internal Stress...................................... 89
        a. Somigliana Dislocations... 89
        b. The Incompatibility Tensor...................................... 91
    5. Elastic Interaction Energies.. 95
        a. Interaction Energies between Stress Systems.................... 95
        b. The Force on a Singularity.. 98
    6. Interaction Energies between Stresses and Inhomogeneities......... 99
    7. The Energy-Momentum Tensor of the Elastic Field.................. 101

ESHELBY, J.D.

III. Applications 107
   8. Point Defects 107
     a. Distortion of Crystals 107
     b. Effect on X-Ray Diffraction 113
     c. Solid Solutions 115
     d. Point Defects in Anisotropic Media 119
     e. Point Defects as Inhomogeneities 122
   9. Dislocations 123
     a. Interaction Energy 123
     b. Image Effects 127
     c. Dislocations in Motion 130
     d. Continuous Distributions of Dislocations 136
  10. Surface and Volume Defects 140

EVANS, A.G.

VOLUME 47, 1994

The Physics and Mechanics of Brittle Matrix Composites

A.G. Evans and F.W. Zok

Materials Department, College of Engineering, University of California, Santa Barbara, California

   I. Introduction 181
     1. Rationale 181
     2. Objectives 183
     3. Approach 191
  II. Interfaces 192
     4. Thermomechanical Representation 192
     5. Measurement Methods 196
     6. Sliding Models 196
     7. Experimental Results 199
     8. Environmental Influences 202
 III. Residual Stresses 203
     9. Origin 203
   10. Measurement Methods 204
   11. Experimental Results 206
 IV. Fiber Properties 207
   12. Load Sharing 207
   13. The Ultimate Tensile Strength 209
   14. Fiber Pull-Out 210
   15. Influence of Flaws 213
   16. *In Situ* Strength Measurements 214
   17. Experimental Results 216

EVA–FAN

EVANS, A.G.

V.	Matrix Cracking in Unidirectional Materials	218
	18. Basic Mechanics	219
	19. The Matrix Cracking Stress	221
	20. Crack Evolution	223
	21. Constitutive Law	226
	22. Simulations	230
	23. Experiments	231
VI.	Matrix Cracking in 2-D Materials	236
	24. Tensile Properties	236
	25. Shear Properties	241
	26. Transverse Tensile Properties	244
VII.	Stress Redistribution	247
	27. Mechanism Transitions	247
	28. Mechanics Methodology	248
	29. Measurements	252
VIII.	Fatigue	255
	30. Basic Phenomena	255
	31. Matrix Crack Growth	259
	32. Multiple Cracking and Failure	264
	33. Thermomechanical Fatigue	266
	34. Experimental Results	268
IV.	Creep	274
	35. Basic Behavior	274
	36. Effect of Fiber Failures	278
	37. Interface Debonding	279
	38. Matrix Cracking	279
	39. Strain Recovery	282
	40. Experimental Results	283
X.	Conclusions	285

FAN, H.Y.

VOLUME 1, 1955

Valence Semiconductors, Germanium and Silicon

H.Y. FAN

Purdue University, Lafayette, Indiana

I. Elements of Semiconductor Theory	284
1. The Perfect Crystal	284
2. Lattice Imperfections	285
3. Electron Distribution	287
4. Extrinsic Semiconductor	288
5. Effective Mass	290

FAN, H.Y.

II. Properties of the Crystal... 291
    6. Crystal Lattice.. 291
    7. Optical Properties and Dielectric Constant....................... 294
    8. Magnetic Properties... 295
    9. Effective Mass and Energy Band Structure....................... 298
    10. Energy Gap.. 303
    11. Impurities... 309
      a. Physical Properties.. 309
      b. Impurity Energy Levels..................................... 311
    12. Lattice Defects.. 315
      a. Thermal Quenching.. 315
      b. Radiation Effect... 316
      c. Dislocations.. 319
III. Conduction Phenomena... 321
    13. General Theory.. 321
    14. Electrical Conductivity and Hall Effect......................... 326
      a. Theory.. 326
      b. Experimental Results....................................... 330
    15. Magnetoresistance... 336
    16. Thermoelectric Power... 340
    17. High Field Conduction.. 345
      a. "Hot" Carriers.. 345
      b. Carrier Multiplication...................................... 347
      c. Zener Current... 348
      d. Ionization of Neutral Impurities............................. 348
    18. Low-Temperature Conduction.................................. 349
    19. High-Frequency Conduction................................... 351
IV. Recombination and Trapping of Carriers............................. 354
    20. Recombination... 354
      a. Recombination Processes.................................... 356
      b. Experimental Results....................................... 358
    21. Trapping Phenomena... 360
      a. Experimental Results....................................... 361

FERREIRA, R.: *see* BASTARD, G.

FER–FRE

FERRY, David K.   VOLUME 49, 1995

Modeling of Quantum Transport in Semiconductor Devices

DAVID K. FERRY

Center for Solid State Electronics, Arizona State University, Tempe, Arizona

HAROLD L. GRUBIN

Scientific Research Associates, Glastonbury, Connecticut

I.   Introduction 283
  1. Quantization in Devices 286
  2. The Difference from Boltzmann Transport 289
  3. Open Systems and Contacts 297
  4. Some Potentially Important Quantum Devices 303
II. The Quantum Equations 308
  5. The Density Matrix and Its Brethren 314
  6. The Kubo Formula and Langevin Equations 327
  7. Boltzmann-like Approaches 334
  8. Moment Equations: The Transition from Classical to Quantum 336
III. Modeling with the Density Matrix 346
  9. Some Considerations on the Density Matrix 346
  10. Dissipation and Current Flow 361
  11. Further Considerations on the Density Matrix 370
IV. Modeling with the Wigner Distribution 378
  12. Methods of Solving the Equations 381
  13. The Double-Barrier Resonant-Tunneling Diode 391
  14. The Role of Dissipation 399
  15. Other Devices 402
V.   Modeling with the Green's Functions 404
  16. Homogeneous, Low-Field Systems 406
  17. Homogeneous, High-Field Systems 417
  18. Femtosecond Laser Excitation 429
  19. The Green–Kubo Formula 435
  20. The Resonant-Tunneling Diode 442

FLORES, F.:   *see* ECHENIQUE, P.M.

FREDERIKSE, H.P.R.:   *see* KAHN, A.H.

FRIEND, Richard H.:   *see* GREENHAM, Neil C.

FULDE, Peter   VOLUME 41, 1988

Theory of Heavy Fermion Systems

PETER FULDE

Max-Planck-Institut für Festkörperforschung,
D-7000 Stuttgart 80, Federal Republic of Germany

JOACHIM KELLER

Fachbereich Physik, Universität Regensburg,
D-8400 Regensburg, Federal Republic of Germany

GERTRUD ZWICKNAGL

Institut für Festköperphysik, TH Darmstadt,
D-6100 Darmstadt, Federal Republic of Germany, and
Max-Planck-Institut für Festkörperforschung,
D-7000 Stuttgart 80, Federal Republic of Germany

I.	Introduction	2
II.	Formation of the Singlet State	7
III.	Quasiparticle Bands	22
	1. Renormalized Band Theory: Formalism	25
	2. Renormalized Band Theory: Results	30
	3. Discussions and Model Descriptions	34
IV.	Quasiparticle–Phonon Interactions	40
	4. Thermodynamic Relations	41
	5. Quasiparticle–Phonon Interaction Hamiltonian.	45
	6. Hydrodynamic Description.	52
V.	Quasiparticle Interactions and Fermi Liquid Description.	63
VI.	Microscopic Theories	71
	7. Expansion Techniques	74
	8. Mean-Field Theory	88
	9. Variational Ground State	98
VII.	Superconductivity	103
	10. Introduction	103
	11. Group Theory.	104
	12. Microscopic Theory.	119
	13. Properties of Different Pair States	137
	Appendix A: Molecular Model for Strongly Correlated Electrons	144
	Appendix B: Parametrization of the Model Hamiltonian	147

GAL–GIL

GALT, J.K.: *see* KITTEL, C.

GEBALLE, Theodore H.: *see* WHITE, Robert M.

GILMAN, J.J.   VOLUME 13, 1962

Dislocations in Lithium Fluoride Crystals

J.J. GILMAN

Brown University, Providence, Rhode Island

AND

W.G. JOHNSTON

General Electric Research Laboratory, Schenectady, New York

I. Introduction.. 148
II. Methods for Observing Dislocations................................ 150
    1. Etching.. 150
    2. Decoration... 158
    3. X-Ray Microscopy... 159
    4. Comparison of Methods....................................... 161
III. Some Properties of Lithium Fluoride Crystals...................... 161
    5. Source.. 161
    6. Dislocations in As-Grown Crystals............................. 162
    7. Glide Systems.. 162
    8. Cross-Glide of Screw Dislocations............................. 164
IV. Behavior of Individual Dislocations............................... 164
    9. Mobility of Freshly Introduced Dislocations.................... 165
    10. A Measurement of Dislocation Line Energy..................... 167
    11. Glide Band Formation from a Dislocation Half-Loop.............. 168
    12. Implications Regarding Crystal Plasticity..................... 168
V. Dislocation Mobilities... 170
    13. Stress Dependence of Dislocation Velocity.................... 170
    14. Temperature Dependence of Dislocation Velocity................ 175
    15. Pinning of Dislocations...................................... 176
VI. Origins of Dislocations.. 177
    16. The First Glide Dislocations................................. 177
    17. Observations of the Multiplication of Glide Dislocations........ 180
    18. The Multiplication Mechanism................................ 185
    19. Growth of Dislocation Structure During Deformation............ 191

GILMAN, J.J.

VII. Theoretical Considerations of Dislocation Mobility 193
    20. Drag of Trails 194
    21. Phonon Viscosity 195
    22. Nucleation Theory 195
    23. The Temperature Dependence 196
    24. Impurity Effects 197
VIII. Individual Dislocations in Strain-Hardened Crystals 197
  IX. Macro-Plasticity 201
    25. Prediction of Stress-Strain Behavior 201
    26. Observed Stress-Strain Behavior 203
    27. Plastic Instability at the Yield Point 205
   X. Effects of Radiation Damage 207
    28. Structural Changes During Irradiation 208
    29. Changes in Dislocation Mobility 209
    30. New Dislocation Sources 211
  XI. Cyclic Strains 211
 XII. Fracture Studies 212
    31. Dislocation Nucleation at Crack Tips 213
    32. Cleavage Step Creation by Dislocations 213
    33. Weakening Caused by Dislocations 214
    34. Surface Energy Measurements 214
XIII. Dielectric Breakdown 216
XIV. Mechanism of Etch-Pit Formation 216
    35. Effect of Dislocations 219
    36. Inhibition of Step Motion 220
    37. Effect of Segregated Impurities 222

GIOVANE, L.: *see* KIMERLING, L.C.

GIVENS, M. Parker   VOLUME 6, 1958

Optical Properties of Metals

M. Parker Givens

University of Rochester, Rochester, New York

  I. Introduction 313
 II. Electromagnetic Theory 316
    1. Relations between the Optical and Electrical Constants 316
    2. The Flow and Conservation of Energy 318
III. Classical Attempts to Predict the Optical Constants 319
    1. The Hagen-Rubens Relation 319
    2. Drude's Theory 320
    3. Short Wavelengths: The Formula of Zener 321
    4. Intermediate Wavelengths 323
    5. $\omega^2 \approx \gamma^2$ 324

GIV–GLI

GIVENS, M. Parker

IV. The Internal Photoelectric Effect.................................... 326
V. The Anomalous Skin Effect... 329
VI. Experimental Methods... 336
   1. Drude's Method... 336
   2. An Alternate Method.. 338
   3. Schulz's Method.. 339
   4. Other Methods... 342
VII. Experimental Results.. 342
   1. General Remarks... 342
   2. Copper, Silver, and Gold...................................... 343
   3. The Alkali Metals.. 348
   4. Divalent Metals 348
   5. Aluminum... 350
   6. The Liquid Metals.. 351
   7. The Anomalous Skin Effect.................................... 351

GLICKSMAN, Maurice   VOLUME 26, 1971

Plasmas in Solids

MAURICE GLICKSMAN

Brown University, Providence, Rhode Island

I. Introduction... 275
   1. History of Plasmas.. 276
   2. Definition of a Plasma 280
   3. Immobile and Mobile Plasmas................................ 289
II. Production of Plasmas.. 291
   4. Thermal Equilibrium.. 291
   5. Injection.. 303
   6. Tunneling and Impact Ionization............................. 308
   7. Irradiation... 317
III. Plasma Properties... 318
   8. DC Conductivity.. 318
   9. High-Frequency Conductivity................................ 322
   10. Properties in an External Magnetic Field..................... 342
   11. Bounded Plasmas... 381
IV. Instabilities.. 386
   12. Classes of Instabilities...................................... 387
   13. Pinch in Self-Magnetic Field................................ 388
   14. Helical Density Wave.. 398
   15. Negative Differential Conductivity and Density Waves......... 410
   16. Two-Stream Instabilities.................................... 415
   17. Other Instabilities.. 421
V. Conclusions.. 426

GOLDBERG, I.B.: *see* WEGER, M.

GOMER, Robert   VOLUME 30, 1975

Chemisorption on Metals

ROBERT GOMER

Chemistry Department and James Franck Institute, University of Chicago, Chicago, Illinois

I.	Introduction	94
II.	A Brief Survey of Chemisorption	94
	1. Summary of Salient Facts	94
	2. Adsorption, Absorption, Reconstruction	96
	3. Comparison of Chemisorption and "Ordinary" Chemical Bonding	97
	4. Topics of Current Interest	97
	5. Where Do We Stand?	98
III.	Experimental Constraints and How to Cope with Them	99
	6. Vacuum Requirements	99
	7. Surface Cleanness and Preparation	102
	8. Surface Characterization	102
IV.	Field Emission and Field Ion Microscopy	104
	9. Theory of Field Emission	104
	10. The Field Emission Microscope	109
	11. Applications of Field Emission Microscopy	113
	12. Field Ion Microscopy	123
	13. Field Desorption	125
V.	Thermal Desorption and Related Phenomena	131
	14. Theory of Thermal Desorption	131
	15. Methods of Studying Thermal Desorption	134
	16. Sticking Coefficients and Coverage Measurements	137
	17. Theory of Sticking Coefficients	139
VI.	Electron Impact Desorption	140
	18. Definitions	140
	19. Theory	143
	20. Applications	150
	21. Some Representative Cross Sections	152
VII.	Theory of Chemisorption	152
	22. Introduction	152
	23. Newns–Anderson Model	154
	24. Generalized LCAO–MO Treatment	161
	25. Adsorbate–Adsorbate Interaction	167
	26. Valence-Bond (Schrieffer–Paulson–Gomer) Approach	168
	27. Linear Response (Kohn–Smith–Ying) Method	169
VIII.	Determination of Electronic Structure of the Adsorption Complex	170
	28. Field Emission Spectroscopy	170
	29. Photoemission	178

GOMER, Robert

30. Auger Methods.. 180
31. Field and Photoemission Distributions from Selected Systems......... 182
32. Adsorbate Charge.. 186
IX. A Look at Some Adsorption Systems.................................... 192
33. CO on Tungsten.. 192
34. Hydrogen on the Tungsten (100) Plane............................. 212
35. Alkali and Alkaline Earth Adsorption............................... 214
X. Note Added in Proof.. 224
36. Very Recent Theoretical Developments.............................. 224

GOURARY, Barry S.   VOLUME 10, 1960

Wave Functions for Electron-Excess Color Centers in Alkali Halide Crystals

BARRY S. GOURARY AND FRANK J. ADRIAN

Applied Physics Laboratory, The Johns Hopkins University, Silver Spring, Maryland

I. Introduction.. 128
1. Some Relevant Properties of the Alkali Halide Crystals.............. 128
2. Scope of Paper.. 130
3. Notation.. 132
II. Summary of the Experimental Situation............................. 134
4. Optical Absorption Spectra and Ivey's Empirical Laws.............. 134
5. Polarization Experiments... 137
6. Inversion Symmetry Determinations................................. 138
7. Electron Spin Resonance and Electron-Nuclear Double Resonance.... 141
III. Empirical Preliminaries.. 143
8. Sizes of Ions... 143
9. Dielectric Constants.. 146
IV. Theoretical Preliminaries.. 148
10. The Quasi-Adiabatic Approximation............................... 148
11. The Hartree-Fock Procedure...................................... 152
12. Comparison... 157
13. Hyperfine Interactions and Their Consequences................... 162
V. Continuum Models... 167
14. Simpson's Treatment of the Interstitial Model of the F Center, and Its Application to the U_2 Center.................................... 168
15. The Work of Pekar and His Collaborators......................... 171
16. The Work of Herman, Wallis, and Wallis on the $F_2{}^+$ and the F_2 Centers 184
17. Pincherle's D-Center Calculation and the Work of Wallis, Herman, and Milnes... 188
VI. Semicontinuum Models.. 188
18. The Jost Cavity Model and the Work of Mott and Littleton......... 189
19. Tibbs' Calculation of the Wave Functions of the F Center in NaCl.... 191
20. The Calculations of Krumhansl and Schwartz, of Krumhansl, and of Dexter... 194
VII. Molecular Orbital Techniques..................................... 203
21. The Molecular Orbital Calculations of Inui and Uemura............ 205
22. The Construction of a Molecular Orbital from Hyperfine Data........ 207

GOURARY, Barry S.

VIII. The Point-Ion-Lattice Approximation and Its Refinements............. 210
    23. The Point-Ion-Lattice Model of the F Center....................... 211
    24. Hyperfine Interactions and g-Factor Shift in the Point-Ion-Lattice Model.. 218
    25. Effects of Exchange and Finite Ion Size in the F Center............. 227
    26. The M Center.. 234
    27. The U Center.. 240
  IX. Conclusions.. 244
   X. Acknowledgments.. 245
      Appendix A. Calculation of Two-Center Integrals Involving the F Electron and the Ion-Core Electrons............................. 245
      Appendix B. Evaluation of the Integral $(\psi_F|g_\alpha|\psi_F)$..................... 246

GREENHAM, Neil C.   VOLUME 49, 1995

Semiconductor Device Physics of Conjugated Polymers

Neil C. Greenham and Richard H. Friend

Cavendish Laboratory, Cambridge, United Kingdom

   I. Introduction.   2
     1. Semiconducting and Conducting Polymers.   2
     2. Organic Semiconductor Devices   7
  II. Materials 12
     3. Synthesis of Semiconducting Polymers 12
     4. Doped Conducting Polymers 25
     5. Molecular Semiconductors 27
     6. Physical Properties of Conjugated Polymers. 29
 III. Optical and Electronic Properties of Conjugated Polymers 32
     7. Introduction. 32
     8. Electronic Structure—Ground State 35
     9. Electronic Structure—Excited States. 39
    10. Experimental Results 51
 IV. Device Processes 63
    11. Charge Injection 64
    12. Carrier Transport. 73
    13. Exciton Decay. 80
  V. Field-Effect Diodes and Transistors 84
    14. The Field-Effect Device 84
    15. Polyacetylene Devices 91
    16. Polythiophene and Poly(p-phenylenevinylene) Devices 103
    17. Thiophene Oligomer Devices 111
    18. Other Devices. 113

GRE–GSC

GREENHAM, Neil C.

VI. Light-Emitting Diodes 115
  19. Background 115
  20. Conjugated Polymer Electroluminescence 117
  21. Control of Color 118
  22. Electrical Characteristics of Single-Layer Devices 124
  23. Multilayer Devices 130
  24. Probing of Excited States in LEDs 134
  25. Optical Properties of LEDs 137
  26. Novel Device Structures 142
  27. Technological Issues 143
VII. Photoconductive and Photovoltaic Devices 144
  28. Molecular Devices 144
  29. Polymeric Devices 145
VIII. Conclusions 149

GRUBIN, Harold L.: *see* FERRY, David K.

GSCHNEIDNER, Karl A., Jr. VOLUME 16, 1964

Physical Properties and Interrelationships of Metallic and Semimetallic Elements

KARL A. GSCHNEIDNER, JR.

Department of Physics and Materials Research Laboratory, University of Illinois, Urbana, Illinois, and University of California, Los Alamos Scientific Laboratory, Los Alamos, New Mexico

  I. Introduction 276
    1. General Description of the Tables of Physical Properties 276
    2. General Description of the Derived Properties 278
  II. Elastic Properties 279
    3. Young's Modulus 280
    4. Shear Modulus 287
    5. Poisson's Ratio 291
    6. Compressibility (Bulk Modulus) 298
  III. Coefficient of Thermal Expansion 313
  IV. Atomic Volume 321
  V. Melting Point and Heat of Fusion 323
    7. International Practical Temperature Scale 323
    8. Melting Point 325
    9. Heat of Fusion 331

GSCHNEIDNER, Karl A., Jr.

VI. Boiling Point and Heat of Sublimation.. 336
 10. Boiling Point.. 336
 11. Heat of Sublimation... 343
 12. Cohesive Energy... 348
VII. Specific Heat Data.. 350
 13. Electronic Specific Heat Constant.. 350
 14. Heat Capacity at Constant Pressure...................................... 362
 15. Heat Capacity at Constant Volume and the Dilation Term............ 365
VIII. Debye Temperature... 368
 16. Debye Temperature from Specific Heat Data.......................... 369
 17. Lindemann Equation and the Debye Temperature................... 378
 18. Debye Temperature from Elastic Constants......................... 379
 19. Debye Temperature from Electrical Resistivity...................... 383
 20. Debye Temperature from Thermal Expansion........................ 383
 21. Debye Temperature from X-Ray Intensity Data.................... 387
IX. Some Interrelationships and Derived Properties......................... 388
 22. Ratio of Young's Modulus to the Shear Modulus.................. 388
 23. The Ratio $(1 - \sigma)/(1 + \sigma)$.. 391
 24. The Ratio of the Compressibility Constants, b/a^2................. 394
 25. The Relationship between Thermal Expansion and Melting Point...... 397
 26. Leibfried, Modified Leibfried, and Bragg Numbers.................. 398
 27. Entropies of Fusion and Vaporization............................... 404
 28. Grüneisen Constant... 410
 29. Size Factor... 419

GUINIER, Andre VOLUME 9, 1959

Heterogeneities in Solid Solutions

ANDRE GUINIER

University of Paris, France

 I. Object of the Article.. 294
 II. Method of Studying Faults in the Periodicity by X-Ray Scattering....... 295
 1. Nature of the Imperfections.. 295
 2. Effect of Imperfections on X-Ray Diffraction........................ 296
 3. General Results of the Theory... 298
 4. General Information on the Nature of Crystalline Disorder Given by
 X-Ray Scattering... 300
 5. Planar and Linear Faults.. 302
 6. Experimental Methods... 305
 III. Solid Solutions in Equilibrium.. 307
 7. Perfect and Imperfect Disorder.. 307
 8. Short-Range Order; Characteristics of the Scattering Produced........ 308
 9. Experimental Evidence of Short-Range Order......................... 309
 10. Quantitative Study of Short-Range Order in Solid Solutions........... 312
 11. Results of Observations on Short-Range Order....................... 314
 12. Displacement Disorder in Solid Solutions............................. 317
 13. Comparison of the Theories of Solid Solutions with the Experimental
 Data.. 321

GUI–GÜN

GUINIER, Andre

IV. Nonequilibrium Solid Solutions... 323
   14. Continuous and Discontinuous Precipitation........................ 324
   15. The Pre-precipitation 325
   16. Zones with or without Distortion.................................. 328
   17. Zones in an Al-Ag Alloy... 329
   18. Zones in an Al-Cu Alloy... 336
      a. X-Ray Scattering Characteristics of the G.P. Zones............... 337
      b. Structure of a G.P. Zone...................................... 342
      c. Disappearance of the G.P. Zones............................... 344
      d. G.P. Zones in Other Systems.................................. 345
   19. Zones Having Linear Disorder..................................... 346
   20. Zones in the Ternary Alloys Al-Zn-Mg............................ 351
   21. Alloys Displaying the Phenomenon of "Side Bands"................. 356
   22. Formation and Stability of Zones................................. 363
   23. Reversion of the Zones... 367
   24. Kinetics of the Formation of Zones............................... 371
      a. Action of Dislocations... 373
      b. Action of Vacancies... 374
   25. Influence of Lattice Faults on the Kinetics of Zone Development....... 378
      a. Influence of Deformation...................................... 378
      b. Influence of Irradiation....................................... 380
      c. Influence of Impurities in the Matrix........................... 382
   26. Properties of the Zones... 384
      a. Changes in Internal Energy.................................... 384
      b. Dimensional Variations of the Crystal.......................... 385
      c. Mechanical Properties... 387
      d. Electrical Properties.. 394
      e. Magnetic Properties... 396
V. Acknowledgments... 398

GUNSHOR, R.L.: *see* NURMIKKO, A.V.

GÜNTHERODT, H.-J.: *see* BUSCH, G.

Order-Disorder Phenomena in Metals

LESTER GUTTMAN

Research Laboratory, General Electric Co., Schenectady, New York

I. Introduction...... 146
II. Structures of Ordered Alloys...... 148
  1. Perfectly Ordered Alloys...... 148
  2. Imperfectly Ordered Alloys; Pair Density Functions...... 148
  3. Short- and Long-Range Order...... 152
  4. The Measurement of the Pair-Density Functions...... 156
  5. Techniques of Measurement of Order Parameters...... 164
  6. Results of Measurement of Short-Range Order...... 166
    a. Single Crystals...... 166
    b. Polycrystalline Specimens...... 168
  7. Results of Measurements of Long-Range Order...... 170
  8. Ordered Structures with Large Periods...... 174
III. Thermodynamic Treatment of Order-Disorder Transitions...... 178
  9. Effect of Temperature on an Ordered Phase...... 178
  10. Phase Transitions in Pure Substances; Higher Degree Transitions...... 179
  11. Phase Transitions in Binary Mixtures; Equilibrium Diagrams...... 184
  12. The Degree of Order-Disorder Transformations...... 186
  13. The Cooperative Nature of Order-Disorder Transformations...... 188
IV. Statistical Thermodynamics of Ordering Alloys...... 191
  14. Survey of Approximate Solutions of the Ising Problem...... 192
  15. Comparison of Approximate Theories with Experimental Results...... 198
    a. Discontinuity in Heat Capacity...... 198
    b. Changes in Heat Content...... 199
    c. Phase Boundaries...... 200
    d. Long-Range Order...... 201
    e. Short-Range Order Parameters and Activity Coefficients...... 201
  16. Appraisal of the Ising Model; Further Developments of Theory...... 204
    a. Neglect of Interaction between Atoms Not Nearest Neighbors...... 205
    b. Neglect of the Changes of Cell Size and Lattice Symmetry or Thermal Expansion during the Order-Disorder Transition...... 206
    c. Neglect of the Effect of Lattice Vibrations on the Ordering...... 208
    d. Limit of Validity of the Ordering Energy Expression...... 209
    e. The Use of Simplified Concepts of Long- and Short-Range Order for Describing the State of Order...... 210
V. The Kinetics of Order-Disorder Transformations...... 210
  17. Resumé of the Theory of Kinetics of Phase Transformations...... 211

GUTTMAN, Lester

18. Kinetics of Order-Disorder Transformations from the Viewpoint of the
    General Theory... 212
    a. Nucleation and Short-Range Order............................ 213
    b. Formation of Out-of-Step Domains........................... 213
VI. Acknowledgments... 216
    Appendix A. Ideal Ordered Structures.......................... 216
    Appendix B. Equilibrium Properties of Ordering Systems: Experimental
    Studies... 220

GUYER, R.A.   VOLUME 23, 1969

The Physics of Quantum Crystals

R.A. GUYER

Department of Physics, Harvard University, Cambridge, Massachusetts

I. Introduction 413
II. Background.. 419
    1. Why Quantum Solids?.. 423
    2. The Hartree Approximation and the Need for Short-Range Correlations. 434
    3. Phonons and the Failure of Conventional Lattice Dynamics........... 439
    4. Experimental Information about Phonons......................... 440
III. Short-Range Correlations.. 444
    5. The Nosanow Cluster Expansion................................ 444
    6. The Brueckner–Frohberg Expansion............................ 456
    7. The Woo-Massey Cluster Expansion: Quasi-Crystal Approximation 461
IV. Phonons.. 462
    8. Self-Consistent Phonons..................................... 462
    9. RPA Phonons... 469
V. A Self-Consistent Theory... 473
    10. A Cluster Expansion Approach................................ 473
    11. Rayleigh–Schrödinger Perturbation Theory 482
VI. Exchange... 485
    12. Calculation of the Exchange Integral.......................... 488
    13. A Critique of the Nosanow Theory of Exchange................. 495
VII. Status of Current Problems in Quantum Solids.................... 496

GUYON, E.:   *see* DUBOIS-VIOLETTE, E.

HAASEN, P.: *see* ALEXANDER, H.

HAHN, E.L.: *see* DAS, T.P.

HALPERIN, B.I. VOLUME 21, 1968

The Excitonic State at the Semiconductor-Semimetal Transition

B.I. HALPERIN AND T.M. RICE

Bell Telephone Laboratories, Incorporated, Murray Hlll, New Jersey

I. Introduction 116
   1. Nature of the Excitonic Phase 116
   2. Phase Diagram 119
   3. Effects of Impurities on the Existence of the Excitonic Phase 124
   4. Choice of Materials 125
   5. Transport Properties 128
   6. Outline of This Article 130
II. Screened Hartree–Fock Approximation 131
   7. Definition of the Approximation 131
   8. Band Notation 135
III. Single Valence Maximum and Single Conduction Minimum 137
   9. General Theory 137
   10. Dominant-Term Approximation 140
   11. Splitting the **M** degeneracy: Case **Q** = **w** 141
   12. Case **Q** ≠ **w** 147
IV. Band Structures with Several Equivalent Electron or Hole Pockets 148
   13. One Band Maximum and Two Equivalent Minima 149
   14. Two Band Maxima and Two Minima 155
V. Region near E_c^+, the Semiconducting Transition Point 156
   15. General Theory 156
   16. Ground State in the Distorted Region 160
   17. Static Linear Response in the Nondistorted State 163
   18. Splitting between the Eigenvalues 165
VI. Dynamic Linear Response Theory 166
VII. The Electron-Phonon Coupling 170
VIII. Summary of Results of the Screened Hartree-Fock Approximation 177
IX. Beyond Hartree-Fock 180
   19. Existence of the Excitonic State 180
   20. Collective Modes 189
Appendix A: Some Symmetry Relations for the Screened Hartree-Fock
   Matrix Elements 190
Appendix B: Choice of Basis Functions 191
Appendix C: Evaluation of the Singlet-Triplet Splitting Constant 192

HAM–HAS

HAM, Frank S.   VOLUME 1, 1955

The Quantum Defect Method

FRANK S. HAM

Department of Physics, University of Illinois, Urbana, Illinois

I. Introduction.. 127
II. The One-Electron Potential and the Hartree-Fock Equation............ 130
III. Fundamental Features of the Quantum Defect Method—the Work of
Kuhn and Van Vleck... 136
IV. Calculation of Energy Bands with QDM............................. 140
V. Proof of the Sufficiency of the Assumptions of QDM.................. 144
VI. The Extrapolated Quantum Defect and the WKB Justification of QDM.. 150
VII. Extrapolation at Low Energies—the "η-Defect"..................... 157
VIII. Corrections for Core Polarization, Overlap, and Valence Electron Inter-
action in Multivalent Atoms....................................... 165
IX. Correlation and Exchange Interaction between Valence and Core Electrons 169
X. Quantum Defect Data for the Alkali Metals......................... 174
XI. Results of Band Calculations with QDM for the Alkali Metals......... 184
XII. Acknowledgment... 190
Appendix: Solutions of the Hydrogenic Wave Equation............... 190

HASHITSUME, Natsuki: *see* KUBO, Ryogo

HASS, K.C.   VOLUME 42, 1989

Electronic Structure of Copper-Oxide Superconductors

K.C. HASS

Research Staff
Ford Motor Company
Dearborn, Michigan

I. Introduction 213
II. Overview of Basic Physical Properties 215
1. $La_{2-x}M_xCuO_{4-y}$ 215
2. $RBa_2Cu_3O_{7-\delta}$ 218
III. One-Electron Band Structures 221
3. Spin-Restricted Calculations 221
4. Spin-Polarized Calculations 233

HASS, K.C.

IV.	Correlated Electron-Electron Models	235
5.	Single-Band Hubbard Hamiltonian	235
6.	Extended Hubbard and Anderson Hamiltonians	239
7.	Role of O p_π Orbitals	248
V.	Implications of Spectroscopic Data	250
8.	Photoemission and Inverse Photoemission	250
9.	Auger	261
10.	X-Ray Absorption and High-Energy Electron Energy Loss	262
11.	Optical and Low-Energy Electron Energy Loss	265
VI.	Concluding Comments	267
	Notes Added in Proof	269

HAYDOCK, Roger VOLUME 35, 1980

The Recursive Solution of the Schrödinger Equation

ROGER HAYDOCK

Cavendish Laboratory, University of Cambridge, Cambridge, England

I.	Basics	216
1.	Introduction	216
2.	Notation	218
3.	The Hamiltonian and Orthogonality	220
4.	The Chain Model	222
5.	Transformation to a Chain	225
6.	A Free Electron	228
7.	Numerical Transformation to a Chain	229
8.	Cluster Size, Chain Length, and Boundary Conditions	231
9.	d Electrons on an fcc Lattice	234
10.	Eigenvalues and Eigenstates of a Chain	236
11.	Irregular Solutions on a Chain	240
12.	The Constant Chain	242
13.	The Chain Model of a Free Electron	243
14.	Symmetry	244
15.	Vector Chains	247
16.	Computation of Eigenvalues and Eigenstates	247
II.	Green Functions and the Local Density of States	250
17.	Green Functions of a Finite Chain	250
18.	Green Functions of an Infinite Chain and Their Integrals	255
19.	Examples of Green Functions of Infinite Chain Models	259

HAY

HAYDOCK, Roger

III. Perturbation Theories, Energy Differences, and Computing the Density of States . 263
  20. Perturbation Theory for the Local Density of States, Van Hove Singularities,
     and Chain Asymptotics 263
  21. Singularities in a Band of Finite Width 270
  22. Computing the Local Densities of States 270
  23. The Average (Total) Density of States 275
  24. Differences in Energy and Other Integrated Quantities 276
  25. Binding Energy 277
  26. Perturbation Theory for the Chain 281
  27. Perturbation of the Green Function and Transition Matrix Elements 283
  28. Perturbation Theory Applied to Integrated Quantities 286
IV. Relation to Other Methods and Conclusion 287
  29. Relation to Other Methods 287
  30. The Method of Moments 289
  31. Transfer Matrices 290
  32. Feenberg Perturbation Theory 291
  33. The Cluster Bethe Method 291
  34. Equation of Motion Method 292
  35. Conclusion 292

HAYES, T.M.

VOLUME 37, 1982

Extended X-Ray Absorption Fine Structure Spectroscopy

T.M. HAYES AND J.B. BOYCE

Xerox Palo Alto Research Center, Palo Alto, California

  I. Introduction 173
    1. Qualitative Description of EXAFS 178
 II. Photoabsorption Theory 182
    2. Photoexcitation Formalism 183
    3. Specialization to EXAFS 185
    4. Underlying Approximations 196
    5. Calculation of EXAFS Spectra 212
    6. Structural Information Content of EXAFS 220
    7. Absorption Edge Features 235
III. Experimental Techniques 237
    8. Transmission Experiments 237
    9. Indirect Methods 253
IV. Data Analysis 274
  10. Data Analysis Techniques 274
 V. Experimental Studies 287
  11. Crystalline Solids 287
  12. Disordered Solids 314
  13. Liquids 328
  14. Surfaces and Adsorbates on Surfaces 333
  15. Catalytic Systems 344
  16. Biological Materials and Chemical Complexes 348

HEBEL L.C., Jr.   VOLUME 15, 1963

Spin Temperature and Nuclear Relaxation in Solids

L.C. HEBEL, JR.

Bell Telephone Laboratories, Incorporated, Murray Hill, New Jersey

I. General Introduction.. 409
II. Basic Spin Temperature Theory....................................... 413
    1. Internal Energy and Magnetization in Semiequilibrium................. 413
    2. Thermodynamics in Semiequilibrium 415
    3. Spin-Lattice Relaxation at Constant Field........................... 418
    4. Spin-Lattice and Cross-Relaxation in Coupled Systems................ 421
III. Spin Temperature and High Magnetic Field........................... 423
    5. Introduction: First-Order Rate Equations............................ 423
    6. Spin Calorimetry.. 425
    7. Ultrasonic Saturation.. 429
    8. Approach to Semiequilibrium in a Quadrupolar Solid................. 431
    9. Spin-Lattice Relaxation by Quadrupolar Coupling.................... 434
IV. Spin Temperature and Low Magnetic Field............................ 437
    10. Introduction: Justification of Semiequilibrium....................... 437
    11. Adiabatic Magnetization... 441
    12. Spin-Lattice Relaxation in Pure Metals............................. 442
    13. Spin Relaxation in Mixed Systems.................................. 448
V. Spin Temperature and Strong RF Field............................... 452
    14. Introduction: Semiequilibrium in the Rotating Frame................. 452
    15. Demagnetization in the Rotating Frame............................. 457
    16. Steady-State Saturation.. 476
    17. Rotary Saturation... 485
    Appendix A.. 486
    Appendix B.. 489

HEDIN, Lars   VOLUME 23, 1969

Effects of Electron-Electron and Electron-Phonon Interactions on the One-Electron States of Solids

LARS HEDIN AND STIG LUNDQVIST

Chalmers University of Technology, Göteborg, Sweden

I. Introduction.. 2
II. The Independent Particle Model..................................... 5
    1. Introduction.. 5
    2. The Hartree–Fock Equations...................................... 6
    3. The Total Energy and the Single Particle Excitation Energies........ 8
    4. Unrestricted Hartree–Fock Theory................................. 10
    5. Time-Dependent Hartree Theory and the Lindhard Dielectric Function 12

HED

HEDIN, Lars

III. Summary of Key Concepts and Formulas in Green Function Theory 19
    6. Introduction 19
    7. Operators and States in Second Quantization 20
    8. Equation of Motion for the Field Operator 22
    9. The One-Electron Green Function at Zero Temperature 25
    10. Spectral Representation of the One-Electron Green Function 26
    11. The Structure of the Spectral Function. Elementary Excitations and
        Quasi Particles 28
    12. The Two-Particle Green Function. Formulas for the Total Energy 31
IV. Equation of Motion for the Electron Green Function 33
    13. Electrons in a Rigid Lattice 34
    14. Physical Interpretation of the Electron Self-Energy 38
    15. Electrons in a Vibrating Lattice 43
    16. The Phonon Contribution to the Electron Self-Energy 48
    17. Remarks on Temperature Green Functions 51
V. The Landau Fermi Liquid Theory 54
    18. Introductory Remarks 54
    19. The Kinetic Equation and Wavelike Excitations 59
VI. Wave-Packet States in Many-Particle Systems 65
    20. Introductory Remarks on Wave Packets 65
    21. The Optical Potential. Scattering on Atoms and Surfaces 68
    22. Motion of Wave Packets in an Extended Medium 69
VII. Properties of an Electron Gas 73
    23. Introduction 73
    24. The Total Energy and the Compressibility 75
    25. The Electron Self-Energy 80
    26. The Fermi Liquid Parameters 96
    27. The Dielectric Function 103
VIII. Phonon Contributions to the Electron Self-Energy 108
    28. Some General Results and Formulas 108
    29. Early Calculations of the Effective Mass. The Work of Engelsberg and
        Schrieffer 111
    30. Calculations with Realistic Phonon Spectra and Including Band-Struc-
        ture Effects 113
    31. Self-Energy and the Spectral Function from Tunneling Data. Density
        of Levels 118
    32. Temperature Effects in Specific Heat and Cyclotron Resonance 120
IX The One-Electron Spectrum in a Real Solid 122
    33. Introductory Remarks about the Self-Energy and the Energy Spectrum 122
    34. The Dielectric Properties 124
    35. The Valence Bands 129
    36. The Core Electrons 135
    37. The Effective Crystal Potential 140
    38. Local Density Approximations 146
X. Discussion of Various Metallic Properties 151
    39. Soft X-Ray Emission Spectra 151
    40. Photoemission and Photoabsorption Spectra 157
    41. Positron Annihilation in Metals 162
    42. Cohesive Energy 168
    43. Fermi Liquid Parameters 171
Appendix A: Linear Response Theory 173
Appendix B: Equations of Motion for Green Functions 175
Appendix C: Spectral Resolutions 180

Localized Moments and Nonmoments in Metals: The Kondo Effect

A.J. HEEGER

Department of Physics and Laboratory for Research on the Structure of Matter
University of Pennsylvania, Philadelphia, Pennsylvania
and
Institut de Physique de la Matière Condensée, Université de Genève, Geneva, Switzerland

I. Introduction.. 284
II. The Formation of Magnetic Moments: Theoretical Discussion 291
   1. The Anderson model... 293
   2. The Admixture Antiferromagnetic s–d Exchange.................... 301
III. Evaluation of the Parameters $\bar{U}$, Δ, and E_d: The Validity of the Anderson
   Model.. 303
   3. The Resonance Energies E_d.................................... 304
   4. The Coulomb Interaction $(\bar{U} + 4\bar{J})$........................... 307
   5. The Virtual Level Width Δ................................... 312
IV. The s–d Model and Experiment................................... 321
   6. Nuclear Magnetic Resonance in Dilute Alloys $(T > T_K)$............. 322
   7. Nuclear Spin Relaxation in Dilute Alloys $(T > T_K)$............... 328
   8. Electron Spin Resonance and Relaxation in Dilute Alloys........... 331
   9. Giant Moments in Exchange Enhanced Metals..................... 339
   10. Summary.. 343
V. The Resistance Minimum Phenomenon: The Kondo Effect.............. 343
   11. The Divergence in Perturbation Theory and Its Implications........ 343
   12. Perturbation Theory of the Magnetic Susceptibility.................. 354
   13. Higher-Order Effects on the RKKY Spin Density Oscillations........ 357
   14. Effect of the Divergence on the Specific Heat $(T \gg T_K)$.............. 358
VI. The Ground State of the Magnetic Impurity Problem: Summary of Theoretical Results for $T < T_K$...................................... 359
   15. The Singlet Ground State Theories................................ 360
   16. The Green's Function and Dispersion Theory...................... 366
VII. The Kondo Effect: Experimental Results for $T < T_K$.................. 372
   17. Transport Properties: Thermoelectric Power and Electrical Resistivity
   $(T < T_K)$.. 375
   18. The Magnetic Susceptibility $(T < T_K)$........................... 382
   19. The Specific Heat $(T < T_K)$................................... 388
   20. Spin Correlations for $T < T_K$; Mössbauer and NMR Studies.......... 392
VIII. Conclusion... 407

HEE

HEER, Ernst   VOLUME 9, 1959

The Interdependence of Solid State Physics and Angular Distribution of Nuclear Radiations

ERNST HEER AND THEODORE B. NOVEY

University of Rochester, Rochester, New York, and Argonne National Laboratory, Lemont, Illinois

I. Introduction... 200
II. The Unperturbed Angular Correlation of Nuclear Radiation........... 202
   1. Angular Correlation of Successive Nuclear Radiations.............. 202
   2. Oriented Nuclei.. 206
   3. Nuclear Reactions.. 209
   4. Decay of Elementary Particles.............................. 210
III. Extranuclear Effects Involving Nonparamagnetic Atoms.............. 211
   5. Introduction... 211
   6. Static External Magnetic Fields............................. 213
   7. Static Electric Quadrupole Interaction....................... 217
   8. Combined Static Magnetic and Axially Symmetric Electric Interaction 225
   9. Time-Dependent Electric Quadrupole Interaction............... 227
   10. Time-Dependent Electric Quadrupole Interaction Plus External Magnetic Field.. 232
   11. Time-Dependent Magnetic Interaction........................ 235
IV. Extranuclear Effects Involving Paramagnetic Atoms................. 235
   12. Introduction.. 235
   13. Free Paramagnetic Atoms.................................. 236
   14. Paramagnetic Atoms in Solutions........................... 237
   15. Paramagnetic Atoms in Solids.............................. 240
V. Extranuclear Fields Created by the Nuclear Process (Radiation Damage) 241
   16. The Effect of the Nuclear Recoil........................... 241
   17. The Effect of Excitation of the Electron Shell................ 242
VI. Special Topics.. 245
   18. The Decay of Elementary Particles.......................... 245
   19. Studies of Very Long-Lived States.......................... 246
VII. Remarks on Experimental Techniques........................... 247
   20. Introduction.. 247
   21. Detectors.. 249
   22. Electronics... 250
   23. Sources.. 252
   24. Corrections... 253
   25. Criteria for Suitable Radioactive Nuclides................... 253
VIII. Angular Distributions of Nuclear Radiations as a Tool in Solid State Physics.. 254

Electronic Processes in Zinc Oxide

G. Heiland and E. Mollwo

Universität Erlangen, Germany

AND

F. Stöckmann

Technische Hochschule Darmstadt, Germany

I. Introduction ... 193
   1. Applications ... 193
   2. Lattice Structure ... 193
   3. Physical Data ... 193
II. Preparation of Samples ... 196
   4. Thin Layers ... 196
   5. Sintered Specimens ... 197
      a. Sintered Layers ... 197
      b. Compact Sintered Specimens ... 198
   6. Single Crystals ... 199
III. Diffusion of Defects ... 203
   7. Zinc ... 203
      a. Thin Layers (Tarnishing Processes with Metallic Zinc) ... 203
      b. Crystals and Sintered Specimens ... 205
   8. Hydrogen ... 209
   9. Other Substances ... 212
   10. Discussion of the Diffusion of Imperfections ... 212
IV. Optical Properties ... 216
   11. Absorption and Refraction ... 216
      a. Materials without Admixtures ... 216
      b. Crystals with Admixtures ... 219
      c. Discussion of the Absorption ... 222
   12. Light Emission ... 223
      a. Thermal Excitation ... 223
      b. Discussion of Thermal Excitation ... 225
      c. Excitation by Photons and Electrons ... 228
      d. Discussion of the Mechanism of Luminescence ... 233
V. Adsorption and Catalysis ... 236
   13. Adsorption ... 237
      a. Oxygen ... 237
      b. Hydrogen ... 237

HEI

HEILAND, G.

14. Catalysis.. 239
    a. The Decomposition of Nitrous Oxide........................... 240
    b. The Oxidation of Carbon Monoxide............................ 241
    c. The Hydrogen-Deuterium Exchange............................ 241
    d. Photocatalytic Properties of ZnO............................. 243
15. Discussion of Adsorption and Catalysis.......................... 244
VI. Conductivity in Thermal Equilibrium............................. 246
16. Methods of Measurement.. 246
17. Thin Layers.. 247
    a. Temperature Dependence of the Conductivity, Hall Effect and Thermoelectric Power... 247
    b. Influence of Additional Foreign Metals on the Conductivity....... 248
18. Sintered Specimens.. 249
    a. The Influence of Oxygen on Conductivity and Thermoelectric Power 249
    b. Temperature Dependence of Conductivity and Hall Effect......... 252
    c. Influence of Added Foreign Oxides on the Conductivity........... 253
19. Single Crystals... 255
    a. Without Arbitrary Additions. Dependence of the Conductivity on the Temperature and Pressure of Oxygen. Thermoelectric Power... 255
    b. Crystals with Additions. Dependence of Conductivity and Hall Effect on Doping and Temperature............................. 259
    c. Surface Conductivity of Single Crystals......................... 268
VII. The Variation of the Conductivity as a Result of Various Influences..... 273
20. Field Effect on Crystals.. 273
21. Influence of Oxygen on the Conductivity of Thin Layers............. 275
22. Variation of the Conductivity by Radiation with Light or Electrons. Slow Processes... 278
    a. Thin Layers.. 278
    b. Sintered Specimens... 282
    c. Single Crystals... 284
23. Variation of the Conductivity by Irradiation with Light or Electrons. Fast Processes... 285
    a. Thin Layers.. 285
    b. Single Crystals... 290
    c. "Electrofax"... 295
VIII. Discussion of the Conductivity (Parts VI and VII)................... 296
24. Volume Effects... 296
    a. Electron Mobility... 296
    b. Volume Conductivity and Thermoelectric Power at High Temperatures... 298
    c. Volume Conductivity and Thermoelectric Power at Low Temperatures... 301
25. Surface Effects... 304
    a. Influence of Boundary Layers upon the Conductance............. 304
    b. Surface Conduction on Single Crystals......................... 308
    c. Field Effect on Single Crystals................................ 312
    d. Influence of Oxygen on the Conductivity of Thin Layers and Sintered Specimens... 313
    e. Irradiation with Light and Electrons. Slow Processes............. 316
    f. Irradiation with Light and Electrons. Fast Processes............. 319

HEINE, Volker: *see* COHEN, Marvin L.

HEINE, Volker VOLUME 24, 1970

The Pseudopotential Concept

VOLKER HEINE

Cavendish Laboratory, Cambridge University, Cambridge, England

I. Introduction... 1
II. Why Plane Waves Do Not Work...................................... 3
III. Augmented and Orthogonalized Plane Waves.......................... 6
IV. The Operator Approach.. 10
V. The Scattering Approach... 14
VI. The Cancellation Theorem... 20
VII. Other Pseudopotentials... 26
VIII. Orthogonality Hole and Nonlocality................................ 28
IX. Sins.. 31
    Appendix. The Linear Chain of Black Boxes.......................... 32

HEINE, Volker VOLUME 24, 1970

Pseudopotential Theory of Cohesion and Structure

VOLKER HEIN AND D. WEAIRE

Cavendish Laboratory, University of Cambridge, Cambridge, England

Introduction.. 250
I. The Field Surveyed... 251
    1. Introduction... 251
    2. Historical Sketch.. 252
    3. Pseudopotentials... 255
    4. Metallic Cohesion and Atomic Volume........................... 259
    5. Pseudopotential Effects in Crystal Structure.................... 266
    6. Balls, Bonds, or Brillouin Zones?............................. 284
    7. Covalent Bonding... 291
    8. Range of Applicability of the Method.......................... 294

HEI

HEINE, Volker

II. General Pseudopotential Theory of Structural Energies 300
    9. Ionic Pseudopotential and Other Ingredients of the Theory 300
   10. Screening, Exchange, and Correlation 305
   11. Nonlocality and the On-Fermi-Sphere Approximation 319
   12. Energy Dependence and Orthogonalization Holes 326
   13. Full Theory for Structure Dependence of Total Energy 331
   14. General Features of the Theory 335
III. Beyond Second-Order Perturbation Theory 336
   15. Validity of Second-Order Perturbation Theory 336
   16. Third-Order Perturbation Theory 345
   17. Strong Perturbations, Phase Shifts, and t-Matrices 346
IV. Covalency 354
   18. Description of Covalency in k Space 354
   19. Dielectric Screening 360
V. Survey of Applications 363
   20. Calculation of Ewald Constants 363
   21. Details of Various Applications including Phonon Dispersion and Stacking Faults 365
   22. Calculations with Pseudopotentials of the Austin–Heine–Sham Family . . 375
   23. Calculations Using the Model Pseudopotential 378
   24. Miscellaneous Other Calculations 380
   25. Selection of Material for Review 383
VI. Monovalent Metals 384
   26. Alkali Metals 384
   27. Noble Metals 388
VII. Divalent Metals 391
   28. Divalent Hexagonal Metals 391
   29. Mercury 401
   30. Alkaline Earths Ca, Ba, and Sr 405
VIII. Trivalent Metals 406
IX. Group IV and V Elements 412
   31. Group IV Semiconductors 412
   32. Tetravalent Metals Sn and Pb 413
   33. Group V Semimetals 415
X. Trends in the Periodic Table 418
   34. Trends in Pseudopotential 418
   35. Trends in Atomic Interaction 420
   36. Trends in Crystal Structure 424
XI. Alloys 427
   37. Introduction 427
   38. Atomic Volume and Interatomic Distance 428
   39. Critique of the Common (Rigid) Band Model 430
   40. Hume-Rothery Alloys and Related Topics 436
   41. Pseudopotential Theory of Alloy Structures 440
   42. Some Applications of Pseudopotential Theory 445
   43. Electronegativity 455
XII. Miscellaneous Topics and Recent Developments 455
   44. Liquid Metals 455
   45. Self-Reciprocal Structures 457
   46. Further Comments on Close-Packed Structures 459
   47. Recent Developments 462
XIII. Conclusion 462

HEINE, Volker   VOLUME 35, 1980

Electronic Structure from the Point of View of the Local Atomic Environment

VOLKER HEINE

Cavendish Laboratory, University of Cambridge, Cambridge, England

I. Fundamental Ideas 1
    1. Throwing Out **k** Space 1
    2. The Invariance Theorem 9
    3. Brief Appendix on the Mathematics of the Green Functions 17
II. The Method of Matching the Green Functions 19
    4. Derivation of the Green Function 19
    5. Boundary Correction Terms in the Invariance Theorem 24
    6. Applications to Alloys, Surfaces. and Molecules 27
III. LCAO: From under a Cloud to Out in the Sun 35
    7. Historical Perspective and Basic Clarifications 35
    8. Electron Correlation in Solids and Molecules 42
    9. The Modern Anderson–Bullett Scheme for the Tight-Binding Method 47
    10. From Atomic Resonances to LCAO Parameters for Transition Metals 57
IV. Calculations with Atomic Orbitals, Atomic Vibrations, and Other Localized Basis
    Sets 69
    11. Working with Clusters, Moments, and Continued Fractions 69
    12. The Haydock Recursion Scheme 87
    13. Total Energy Calculations 92
    14. Various Types of Perturbation 99
V. Miscellaneous Properties from the Local Point of View 105
    15. A Local Computational Scheme for the Density Functional Formalism ... 105
    16. Pressure, Force, and Energy 110
    17. Self-Consistency in the Potential 121
VI. Historical Note 123
    18. The Development of the Concepts of Local Electronic Structure 123

The Electron–Hole Liquid in Semiconductors: Experimental Aspects

J.C. HENSEL, T.G. PHILLIPS, AND G.A. THOMAS

Bell Laboratories, Murray Hill, New Jersey

I.　Introduction 88
II.　Macroscopic Properties of the e–h Condensate 94
　　1.　Light-Scattering Methodology 94
　　　　a.　Scattering 95
　　　　b.　Transmission 96
　　　　c.　Optical Indices of e–h Drops 97
　　2.　The Size of e–h Drops 100
　　　　a.　Germanium 100
　　　　b.　Silicon 109
　　3.　Steady-State Distribution of Drops 110
　　　　a.　Diffusion Models 110
　　　　b.　Medium Excitation Intensity: Stripe Geometry 112
　　　　c.　High Excitation Intensity: Point Geometry 112
　　　　d.　Dynamical Cloud Models 116
　　4.　Concentration of Drops 120
　　　　a.　Volume Excitation 120
　　　　b.　Surface Excitation 121
　　5.　Formation of Drops 123
　　　　a.　Nucleation 123
　　　　b.　Onsets 132
　　6.　Surface Properties of e–h Drops 137
　　　　a.　Surface Energy: $\mathscr{S}$ 137
　　　　b.　Net Charge: Q 140
　　7.　Motion of Drops 143
　　　　a.　Transport by Diffusion 143
　　　　b.　Transport by Drift 148
　　　　c.　Damping of Droplet Motion 154

HENSEL, J.C.

III.	Time-Dependent Phenomena and Decay of Drops	159
	8. dc Photoconductivity	160
	9. Decay Kinetics of the Droplet System	163
	a. The Kinetic Equations and Their Solutions	163
	b. Microwave Cyclotron Resonance	166
	c. Luminescence Decay Kinetics	173
	d. Microwave Electric Susceptibility	182
	e. Far Infrared Absorption	185
	f. Decay Kinetics under the Influence of Large Perturbations	186
	10. Microwave Breakdown	196
	11. Microwave Dimensional Resonances	199
	12. Phonon Generation and Heating Effects	200
	13. Light-Scattering Kinetics	202
IV.	Experimental Characterization of the Metallic Ground State at $T = 0$	203
	14. Ground-State Energy	204
	a. The e–h Liquid Luminescence Lineshape and Characteristic Energy: μ'	207
	b. The Exciton Lineshape and Spectral Energy: ϵ_x'	214
	c. The e–h Liquid Binding Energy: ϕ	219
	d. The Exciton Binding Energy: ϵ_x	222
	e. Ground-State Energy: E_G	225
	15. Ground-State e–h Pair Density: n_0	226
	a. Luminescence Natural Linewidth: $E_F(0)$	226
	b. Magneto-oscillatory Effects	229
	c. The Plasma Frequency: Resonant Absorption in the Far Infrared	231
	16. The Enhancement Factor: $g_{eh}(0)$	236
	17. Band Structure in the e–h Liquid: Mass Renormalization	238
	a. Magnetoeffects in the Luminescence Lineshape	239
	b. Magnetoplasma Resonances	239
	18. Perturbations on the Ground State	242
	a. Uniform Uniaxial Stress	243
	b. Magnetic Fields	253
	c. Impurities and Alloys	265
V.	Experiments on the Phase Diagram of the Electron–Hole Fluid	280
	19. The Liquid at Low Temperatures	280
	a. Thermal Expansion	280
	b. The Chemical Potential	287
	20. The Exciton Gas at Low Temperatures	289
	a. Onset Measurements	290
	b. Spectroscopic Measurements	295
	21. Completion of the Phase Diagram	297
	a. The Liquid–Gas Critical Point	297
	b. Metal–Insulator Transition	307
VI.	Epilog	312

HER

HERZFELD, Charles M. VOLUME 12, 1961

Group Theory and Crystal Field Theory

CHARLES M. HERZFELD

National Bureau of Standards

AND

PAUL H.E. MEIJER

Catholic University of America

I. Introduction.. 2
II. The Physical Idea of Crystal Field Theory............................. 3
III. Introduction to Group Theory.. 9
    1. General Remarks.. 9
    2. Finite Groups... 11
    3. Continuous Groups... 13
    4. Representation Theory. Elementary................................ 14
    5. Representation Theory. Advanced.................................. 18
    6. Representations of Continuous Groups. Rotation Group............. 23
    7. Survey of the Crystallographic Point Groups...................... 27
    8. Double Groups... 31
    9. Some Remarks on the Reality of Irreducible Representations of Finite
        Groups.. 34
IV. The Uses of Group Theory in Quantum Mechanics........................ 37
    10. The General Connections... 37
    11. The Role of Schur's Lemmas in Quantum Mechanics................. 41
    12. The Wigner-Eckart Theorem....................................... 44
    13. The Pauli Exclusion Principle................................... 47
    14. External Perturbations and Selection Rules...................... 49
    15. General Tensor Coupling... 52
    16. Time Reversal and Spin. Kramers' Theorem........................ 55
    17. Generators and Quantum Numbers.................................. 57
    18. The Jahn-Teller Theorem... 59
V. Crystal Field Theory... 61
    19. The Forms of the Potential Energy............................... 61
    20. The Hamiltonian... 63
    21. The Spin Hamiltonian (Transition Metal Ions).................... 66
    22. The Angular Momentum Hamiltonian (Rare Earth Ions).............. 72
    23. The Calculation of Wave Functions............................... 76
    24. The Maximum Simplification of the Hamiltonian................... 80
Appendix A. The Group Algebraic Calculation of Wave Functions.............. 81
Appendix B. Systematic Index of Group Data................................ 87

HIDESHIMA, T.: *see* SAITO, N.

HIRTH, J.P.: *see* POND, R.C.

HUEBENER, R.P. VOLUME 27, 1972

Thermoelectricity in Metals and Alloys

R.P. HUEBENER

Argonne National Laboratory, Argonne, Illinois

I. Introduction	64
II. Seebeck, Peltier, and Thomson	65
III. Electron Diffusion	69
1. Thomson Coefficient and the Specific Heat of the Conduction Electrons	69
2. Nonequilibrium in the Electron System	70
3. Two Conduction Bands	74
4. Two Scattering Mechanisms: Influence of Impurities	75
IV. Phonon Drag	80
5. Nonequilibrium in the Phonon System	80
6. Effect of Phonon Drag on the Electrical Resistivity	86
7. Influence of Impurities	86
8. Separation of the Components S^e and S^g	87
9. Low Temperature Anomaly and Anisotropy in the Electron Scattering	89
V. Pure Metals	91
10. Compilation of Thermopower Measurements	91
11. Monovalent Metals	96
12. Polyvalent Metals	102
13. Transition Metals	106
14. Lanthanides and Actinides	109
VI. Alloys	109
15. Electron Diffusion Thermopower of Alloys	110
16. Phonon Drag Thermopower of Alloys	118
VII. Size Effect	120
VIII. Superconductors	123
17. Motion of the Magnetic Flux Structure	123
18. Forces on Flux Lines and Entropy Transport	125
19. Experiments and Comparison with Theory	130

HUI, P.M.   VOLUME 49, 1995

Photonic Band-Gap Materials

P.M. HUI

Department of Physics, The Chinese University of Hong Kong, Shatin, New Territories, Hong Kong

NEIL F. JOHNSON

Department of Physics, Clarendon Laboratory, University of Oxford, Oxford OX1 3PU, England, United Kingdom

I.	Introduction	151
	1. Survey	151
II.	Experimental Overview	157
	2. Experimental Systems and Results	157
	3. Applications	165
III.	Theoretical Background	170
	4. Photonic Band Theory: Formalism	170
	5. Theoretical Calculations and Results	186
IV.	Conclusions and Future Direction	199

HUNTINGTON, H.B.   VOLUME 7, 1958

The Elastic Constants of Crystals

H.B. HUNTINGTON

Rensselaer Polytechnic Institute, Troy, New York

I.	General Considerations	214
	1. Introduction	214
	2. Stresses and Strains	216
	3. The Moduli of Elasticity and Compliance	219
	a. Traditional Elasticity	219
	b. Tensor Notation and Equations of Motion	222
	c. The Definition and Meaning of the Elastic Constants	224
	4. Atomistic Theories of Elastic Constants	226
	5. Other Tensor Properties	243
	6. Interrelation with Other Physical Studies	252

HUNTINGTON, H.B.

II. Experimental Methods.. 256
   7. Static Measurements... 257
   8. Dynamic Measurements... 259
   9. Methods Involving Lattice Interaction with Radiation............... 272
III. Elastic Constants for Various Materials............................. 273
  10. Alkali Halides.. 279
  11. Monovalent Metals.. 287
  12. Multivalent Metals.. 291
    a. Aluminum.. 291
    b. Indium.. 294
    c. Beta-Brass.. 294
    d. Hexagonal Metals...................................... 296
  13. Crystals with Diamond or Zincblende Lattice................... 299
  14. Rare Gas Solids.. 303
  15. Ferromagnetic Materials, the ΔE Effect.................. 305
  16. Piezoelectric Materials...................................... 310
  17. Polycrystalline Materials.................................... 316
IV. Variation with Temperature and Pressure........................... 320
  18. Résumé of Experimental Results.............................. 320
  19. Theory of the Equation of State.............................. 325
 V. Influence of Composition, Phase Changes, and Relaxation Phenomena.... 330
  20. Anelastic Effects... 330
  21. Certain Alloy Systems and Mixtures.......................... 332
  22. Influence of Dilute Alloying................................. 335
  23. Effect of Phase Changes on Elastic Constants................. 338
  24. Effect of Superconductivity................................. 341
  25. Effect of Dislocations....................................... 342
  26. Effect of Radiation Damage.................................. 345
  27. Electron Relaxation at Low Temperature...................... 347
VI. Acknowledgments.. 348
VII. List of Symbols.. 349

HUT–INO

HUTCHINGS, M.T. VOLUME 16, 1964

Point-Charge Calculations of Energy Levels of Magnetic Ions in Crystalline Electric Fields

M.T. HUTCHINGS

*The Clarendon Laboratory, Oxford, England and
Yale University, New Haven, Connecticut*

I.	Introduction	227
II.	Determination of the Perturbing Hamiltonian	228
	1. Evaluation of the Crystalline Electric Potential near a Magnetic Ion in Terms of Cartesian Coordinates	229
	2. Evaluation of the Crystalline Electric Potential near a Magnetic Ion in Terms of Spherical Harmonics	237
	3. Number of Harmonics Occurring in the Expression for the Potential	246
III.	Calculation of the Matrix Elements of the Crystalline Potential Perturbing Hamiltonian	247
	4. Direct Integration	248
	5. Use of Stevens' "Operator Equivalents" Method	249
	6. Operator Equivalents for Cubic Potentials	265
IV.	Summary	271
	Bibliography	272

INOKUCHI, Hiroo VOLUME 12, 1961

Electrical Conductivity of Organic Semiconductors

HIROO INOKUCHI AND HIDEO AKAMATU

Department of Chemistry, The University of Tokyo, Tokyo

I.	Introduction	93
	1. The Historical Survey of Electrical Conduction in Organic Compounds.	94
	2. The Concept of Organic Semiconductors	96
	3. The Classes of the Organic Semiconductors	98
	4. The Structural Aspect of the Organic Semiconductors	99
II.	Preparations and Purifications	102
	5. The Practice of Preparations	102
	6. The Purifications of Organic Semiconductors	103
III.	Electrical Resistivity of Organic Semiconductors	107
	7. The Preparation of Conductivity Cells	107
	8. The Survey of the Resistivity Measurements	113
	9. The Pressure Dependence for Resistivity	118
	10. The Single Crystals	120
	11. The Results with Thin Films	123
	12. The Phthalocyanines	125

INOKUCHI, Hiroo

IV. Photoelectric Phenomena 127
   13. The Photoconductivity 128
   14. The Photovoltaic Effect in Organic Semiconductors 135
V. Discussion of the Energy Gap 136
   15. The Energy Band Model 137
   16. The Tunnel Effect 137
   17. The Exciton 138
VI. Semiconductive Properties of Aromatic Complexes 140
   18. The Aromatic-Halogen Molecular Complexes 140
   19. The Alkali Metal-Aromatic Complexes 143
Appendix I. The Molecular Formulas of Semiconductive Organic Compounds . . . 146
   20. The Polycyclic Aza-Aromatic and Aromatic Compounds 146
   21. The Phthalocyanines and the Dyes 147
Appendix II. The Commercial Names of Vat Dyestuffs 148

IPATOVA, I.P.:   *see* MARADUDIN, A.A.

ISIHARA, A.   VOLUME 42, 1989

Electron Correlations in Two Dimensions

A. Isihara

Department of Physics
State University of New York at Buffalo
Buffalo, New York

Preface 271
I. Ground-State Energy 272
   1. Introduction 272
   2. Dielectric Function 277
   3. Plasmon Dispersion 283
   4. Ground-State Energy and Specific Heat 288
II. Phase Transitions 293
   5. Valley Occupancy Phase Transition 293
   6. Gas and Liquid Phases 296
   7. Wigner Lattice 301
   8. Melting of a Wigner Crystal 308

ISI–JAM

ISIHARA, A.

III. Correlations in a Magnetic Field 314
    9. Effective g-Factor 314
   10. De Haas–Van Alpen and Magnetothermal Effects 320
   11. Magnetoconductivity 330
   12. Cyclotron Resonance 339
   13. Localization 349
IV. Quantized Hall Effect 358
   14. Integral Qunatized Hall Effect 358
   15. Interpretation of the IQHE 365
   16. Fractional Quantized Hall Effect 373
   17. $v = 1/m$ Fractional State 378
   18. Hierarchy of Fractional States 387
   19. Collective Excitations and Closing Remarks 391
   Appendix   Glossary of Symbols 401

IWAYANAGI, S.: *see* SAITO, N.

JAMES, R.W.   VOLUME 15, 1963

The Dynamical Theory of X-Ray Diffraction

R.W. JAMES

Physics Department, Universityof Cape Town, South Africa

I. Introduction 55
II. The Geometrical Theory 61
  1. Introduction 61
  2. The Transform Considered as a Distribution in Reciprocal Space 64
  3. The Essential Problem in Structure Determination 65
  4. Periodic Structures 66
  5. The Reciprocal Lattice 68
  6. The Transform of a Crystal Block 68
  7. The Electron Density Expressed as a Fourier Series 69
  8. Discussion 70
  9. Lattice Functions 71

JAMES, R.W.

III. The Electromagnetic Basis of the Theory.............................. 72
  10. Introduction... 72
  11. The Effective Dielectric Constant of the Crystal.................... 73
  12. The Dielectric Constant and its Fourier Coefficients................ 77
  13. The Wave Equation and its Solution............................. 78
  14. The Magnetic Field... 80
  15. The Resonance Error.. 81
IV. The Dispersion Surface.. 82
  16. The Definition of the Surface................................... 82
  17. The Case of a Single Wave Train................................ 84
  18. The Solution for Two Effective Reciprocal-lattice Points $(n = 2)$...... 85
  19. The Nature of the Dispersion Surface for $n = 2$................... 86
  20. The Relative Amplitudes and Phases of the Waves when $n = 2$....... 89
  21. The Transition from Two Waves to One.......................... 89
  22. The Case of Three Waves...................................... 90
  23. Some Remarks on the Solution for the Infinite Crystal.............. 92
V. The Field in a Bounded Crystal..................................... 92
  24. Introduction... 92
  25. The Cases of Transmission and Reflection........................ 94
  26. The Determination of the Wave Points........................... 95
  27. Some Necessary Geometrical Results............................. 98
  28. The Accommodation and the Resonance Errors.................... 101
  29. The Calculation of the Accommodation g........................ 102
  30. A Summary of Formulas.. 103
VI. The Thick Crystal with Negligible Absorption....................... 105
  31. Introduction... 105
  32. Case I (Transmission): Some Geometrical Results.................. 106
  33. The Boundary Conditions and Wave Amplitudes.................... 108
  34. Discussion of the Fields in Case I............................... 110
  35. Reflection from a Thick Nonabsorbing Crystal..................... 112
  36. The Boundary Conditions....................................... 114
  37. The Reflection Coefficient...................................... 115
  38. The Nature of the Wave Field when g is Complex. Primary Extinction.. 116
  39. The Magnitude of the Primary Extinction......................... 118
  40. The Range of Total Reflection and the Deviation from Bragg's Law.... 118
  41. The Integrated Reflection....................................... 121
  42. Secondary Extinction.. 123
  43. The Nature of the Wave Field in a Thick Nonabsorbing Crystal...... 124
  44. The Field on the Atomic Planes................................. 126
  45. The Nature of the Stationary Wave Field in the Crystal............. 129
  46. The Oscillatory or Pendulum Solution in Case I................... 130
VII. The Wave Field in Crystals with Finite Absorption.................. 132
  47. Preliminary Considerations..................................... 132
  48. Crystal without Symmetry Center............................... 134
  49. The Relation of the Absorption Coefficient to the Complex Amplitude.. 136
  50. The Absorption of the Wave Fields.............................. 137
  51. The Calculation of the Absorption Coefficients.................... 138
  52. Absorption in Case I (Transmission)............................. 140
  53. Absorption in Case II (Reflection).............................. 140
  54. The Relation between the Complex Parameters β, p, and v........... 143

JAM

JAMES, R.W.

55. Discussion of the Absorption in Case I (Transmission)............... 144
56. The Asymmetry of the Absorption in Case I...................... 145
57. Discussion of the Absorption in Case II (Reflection)................ 148
VIII. The Reflection and Transmission Coefficients for Crystals of Finite Thickness. 153
58. The Boundary Conditions....................................... 153
59. Case I. Transmission through an Absorbing Crystal................. 154
60. Crystal Slice with Negligible Absorption.......................... 157
61. The Fine Structure of $R_t(p)$..................................... 157
62. The Reflection Coefficient for a Very Thin Slice.................... 160
63. Numerical Discussion of the Formulas for Absorbing Crystals......... 160
64. The Variation of R_t and $\bar{T}$ with the Thickness of the Slice........... 162
65. Reflection and Transmission Coefficients for a Slice of Finite Thickness
in Case II (Reflection)... 165
66. Crystal Slice with Negligible Absorption.......................... 167
67. Discussion of the Formulas for the Nonabsorbing Crystal............ 168
68. Reflection Coefficient for a Thick Absorbing Crystal................ 171
IX. The Propagation of Energy in the Wave Field......................... 175
69. Introduction.. 175
70. The Mean Energy Current...................................... 176
71. Two Relevant Reciprocal-lattice Points........................... 181
72. The Energy Current-density as a Function of the Angle of Incidence... 182
73. Crystals with Negligible Absorption. The Pendulum Solution.......... 183
74. The Energy Current-densities in the Fields Associated with Individual
Wave Points.. 185
75. The Relation of the Direction of Energy Flow to the Dispersion Surface.. 189
76. The Resultant Energy Current-density and the Pendulum Solution..... 192
77. The Lines of Energy Flow...................................... 194
78. The Effect of Absorption on the Energy Current.................... 199
79. The Nature of the Energy Flow-lines in an Absorbing Crystal......... 200
80. The Periodic Component of the Energy Current in Absorbing Crystals.. 202
81. The Energy Current in Thick Absorbing Crystals................... 205
X. Some Experimental Consequences of the Dynamical Theory.............. 207
82. Introduction.. 207
83. The Propagation of Beams of Limited Width...................... 207
84. Limited Beams in Absorbing Crystals............................ 209
85. The Influence of Crystal Distortion on the Energy Propagation........ 211
86. Experimental Proof of Abnormal Transmission in Perfect Crystals 212
87. Transmission of Radiation through Wedge-shaped Crystals and the Experimental Demonstration of the Pendulum Effect................. 213

JAN, J.-P.   VOLUME 5, 1957

Galvanomagnetic and Thermomagnetic Effects in Metals

J.-P. JAN

Laboratoire Suisse de Recherches Horlogères, Neuchâtel, Switzerland

1. Introduction.. 2
I. Phenomenological Theory.. 3
   2. Enumeration of Independent Effects in an Isotropic Medium......... 3
   3. General Definition of the Effects................................ 4
   4. Isothermal and Adiabatic Effects................................ 9
   5. Relationships between the Effects................................ 11
   6. Formulation of the Effects in Single Crystals.................... 11
II. Magnetoresistance and Hall Effect.................................. 15
   7. Generalities.. 15
   8. Experimental Considerations Concerning the Hall Effect.......... 17
      a. Measurable Hall Effect...................................... 17
      b. Corrections Due to the Finite Length of the Hall Plate...... 19
      c. Apparatus and Methods...................................... 21
   9. Theories of Magnetoresistance................................... 24
      a. The Two-Band Model... 27
      b. The Justi-Kohler Diagram................................... 29
   10. Experimental Results on Magnetoresistance...................... 30
      a. Large Magnetic Fields...................................... 31
      b. Anomalous Magnetoresistance and Resistance Minimum.......... 34
      c. References on Magnetoresistance............................. 37
   11. Theory of the Hall Effect...................................... 38
      a. The Two-Band Model... 39
   12. Experimental Results on Hall Effect............................ 41
   13. Hall Effect and Magnetoresistance in Alloys.................... 46
   14. Galvanomagnetic Effects at Very High Frequencies............... 50
   15. Galvanomagnetic Effects in Liquid Metals....................... 52
   16. Hall Effect in Superconductors................................. 53
III. Thermal Conductivity in a Magnetic Field.......................... 54
   17. Theory... 54
   18. Experimental Results on Thermal Conductivity in a Magnetic Field.... 56
IV. Other Effects... 61
   19. Change in Thermoelectric Power in a Magnetic Field............. 61
   20. The Four Transverse Effects.................................... 64

JAN, J.-P.

V. Galvanomagnetic and Thermomagnetic Effects in Ferromagnetics 66
  21. Introduction 66
  22. Magnetoresistance in Ferromagnetics 67
    a. Survey of the Experimental Results 70
  23. Hall Effect in Ferromagnetics 74
    a. Survey of the Experimental Results 78
  24. Change in Thermoelectric Power of Ferromagnetics 82
  25. Other Effects and Miscellaneous 84
VI. Special Effects 85
  26. Galvanomagnetic and Thermomagnetic Effects in Bismuth 85
  27. Oscillatory Galvanomagnetic and Thermomagnetic Effects 87
  28. Effects of Elastic or Plastic Strain 90
  29. Thin Films and Wires 91
  30. Practical Applications 95
  31. Units and Conversion Factors 95

JANSSEN, T.:   *see* CURRAT, R.

JARRETT, H.S.   VOLUME 14, 1963

Electron Spin Resonance Spectroscopy in Molecular Solids

H.S. JARRETT

E. I. duPont de Nemours and Company, Wilmington, Delaware

  I. Introduction 215
 II. The Hamiltonian 218
    1. Introduction 218
    2. Relativistic Formulation 219
    3. Nonrelativistic Limit 222
III. Strong Electrostatic Interactions 224
    4. General 224
    5. Ligand Field Theory 234
    6. Covalent Bonding 242
    7. Spin-Orbit Interaction 256
    8. Spin Hamiltonian 271
 IV. Weak Interactions 275
    9. General 275
   10. Resonance Absorption 276
   11. Hyperfine Structure 295
  V. Electron Spin Resonance of Excited States 334
   12. Organic Molecules 335
   13. Coupled Metal Ions 339

JOANNOPOULOS, J.D.   VOLUME 31, 1976

Theory of Short-Range Order and Disorder in Tetrahedrally Bonded Semiconductors

J.D. JOANNOPOULOS

Department of Physics, Massachusetts Institute of Technology, Cambridge, Massachusetts

AND

MARVIN L. COHEN

Department of Physics, University of California, Berkeley, California, and Inorganic Materials Research Division, Lawrence Berkeley Laboratory, Berkeley, California

Introduction	71
I. "The Amorphous Problem"	72
1. Crystals and Noncrystals	72
2. Near the Fundamental Gap	75
3. Away from the Fundamental Gap	76
II. Methods to Study SRO and SRD	83
4. Hamiltonians	83
5. Complex Crystalline Models	90
6. "Cluster-Bethe Lattice" Models	96
III. Results for Homopolar Systems	100
7. Electronic Densities of States	101
8. Optical Properties	118
IV. Results for Heteropolar Systems	122
9. A Study of Disorder (U)	123
10. A Study of Disorder (L)	130
V. Final Remarks	148

JOHNSON, Neil F.:   *see* HUI, P.M.

JOHNSTON, W.G.:   *see* GILMAN, J.J.

JØR

JØRGENSEN, Chr. Klixbüll VOLUME 13, 1962

Chemical Bonding Inferred from Visible and Ultraviolet Absorption Spectra

CHR. KLIXBÜLL JØRGENSEN

Cyanamid European Research Institute, Cologny (Geneva), Switzerland

I. Electron Configurations.. 376
   1. Atomic Spectroscopy (Spherical Symmetry)......................... 376
   2. Molecular Orbital (MO) Theory..................................... 378
   3. LCAO Approximation... 384
   4. Expanded Radial Function Model................................... 386
II. Interelectronic Repulsion Effects...................................... 388
   5. Multiplet Term-Splitting of a Pure Configuration.................. 388
   6. Correlation Effects.. 398
III. Types of Experimental Data.. 418
   7. Spectrochemical Series... 418
   8. Nephelauxetic Series... 425
   9. $nl \rightarrow n'l'$ Transitions.................................. 427
   10. Electron Transfer Spectra in Complexes........................... 431
   11. Alkali Halide Spectra.. 438
   12. X-Ray Spectra.. 442
IV. Critical Evaluation of Modern Theories................................. 445
   13. Difficulties in the Valence-Bond and Hybridization Descriptions.. 445
   14. Difficulties for the Energy Band Description..................... 448
   15. Difficulties for the Electrostatic Ligand Field Model............ 450
   16. MO Energy Differences and Electronegativities................... 452
   17. Attempts to Define Absolute MO Ionization Energies.............. 455
   18. MO Interelectronic Repulsion Parameters......................... 456
V. The Understanding of Chemical Bonding.................................. 457
   19. Fractional Charges on Atoms in Molecules........................ 457
   20. Classification Rather Than *a priori* Calculation............... 459
   Acknowledgments.. 462

JOSHI, S.K.   VOLUME 22, 1968

Lattice Dynamics of Metals

S.K. JOSHI AND A.K. RAJAGOPAL

Physics Department, University of California, Riverside, California

1. Introduction... 160
2. The Adiabatic Approximation... 161
3. The Classical Theory of Lattice Vibrations in Harmonic Approximation...... 164
4. Experimental Study of Lattice Vibrations in Metals....................... 170
   A. Diffuse X-ray Scattering Method...................................... 170
   B. Coherent Inelastic Neutron Scattering Method........................ 171
   C. Incoherent Inelastic Scattering of Neutrons......................... 173
5. Frequency Distribution Function... 173
6. A Few Oversimplified Models for the Lattice Dynamics of Metals.......... 175
7. Force Constant Models... 176
   A. Form of Force Constant Matrices..................................... 176
   B. Method of Long Waves.. 178
   C. General Discussion of Various Specific Force Model Calculations for Metals 179
   D. Axially Symmetric Model... 186
8. Interpretation and Analysis of the Experimental Normal Mode Frequencies.. 190
   A. Fourier Series Analysis of Dispersion Curves........................ 190
   B. Force-Constant Analyses of the Experimentally Determined Dispersion
      Curves for Various Metals... 193
   C. Use of a Sum Rule for the Study of the Nature of Interatomic Forces..... 200
   D. Interatomic Potential from an Analysis of Phonon Frequency Measurements 202
9. Phenomenological Models... 204
10. Fundamental Theories... 211
    A. Basic Methods and Approximations................................... 212
    B. General Survey of the Calculations of the Electron-Ion Matrix Element... 225
    C. Specific Calculations of Phonon Frequencies in Metals.............. 237
11. Field Theoretic Treatment of Electron-Phonon Interaction and Lattice Dy-
    namics of Metals.. 248
    A. Mathematical Preliminaries... 250
    B. General Formulation of the Problem................................. 254
    C. Some Approximate Schemes... 263
    D. Kohn Anomaly... 284
12. Epilogue... 293
    Appendix A... 295
    Appendix B... 299
    Appendix C... 306
    Appendix D... 309

Oscillatory Behavior of Magnetic Susceptibility and Electronic Conductivity

A.H. KAHN AND H.P.R. FREDERIKSE

Solid State Physics Section
National Bureau of Standards, Washington, D.C.

KÄNZIG, Werner   VOLUME 4, 1957

Ferroelectrics and Antiferroelectrics

WERNER KÄNZIG

General Electric Research Laboratory, Schenectady, New York

I. Introduction... 5
1. Definition of a Ferroelectric, Outstanding Properties................ 5
2. Phenomenological Classifications of Ferroelectrics................... 6
3. Antiferroelectrics, Ferrielectrics.................................. 7
4. History.. 8
5. The Present Understanding of Ferroelectric and Antiferroelectric
Phenomena... 10
II. The Phenomenological Behavior of Ferroelectrics: Experimental......... 11
6. The Dielectric Susceptibility of Ferroelectrics..................... 12
   a. General Dielectric Behavior.................................... 12
   b. KH_2PO_4-Type Ferroelectrics................................. 13
   c. Rochelle Salt.. 17
   d. Other Ferroelectric Tartrates................................. 20
   e. Guanidine Aluminum Sulfate Hexahydrate....................... 21
   f. Ferroelectric Alums.. 21
   g. $BaTiO_3$-Type Ferroelectrics................................. 21
   h. Lithium Niobate and Lithium Tantalate........................ 26
   i. Pyrochlore-Type Ferroelectrics............................... 27
   j. Lead Metaniobate, $PbNb_2O_6$................................ 28
   k. Ferroelectric Sulfates....................................... 28
   l. Recent Discoveries... 29
7. Dielectric Nonlinearity and Saturation............................ 29
   a. General Remarks.. 29
   b. Dielectric Saturation in KH_2PO_4 above the Curie Point......... 29
   c. Dielectric Saturation in Rochelle Salt above the Upper Curie Point 32
   d. Nonlinear Dielectric Behavior of $BaTiO_3$ in the Paraelectric Phase. 32
8. The Electromechanical Behavior................................... 33
   a. Survey... 33
   b. KH_2PO_4-Type Ferroelectrics................................. 34
   c. Rochelle Salt.. 42
   d. Lithium-Ammonium Tartrate Monohydrate..................... 46
   e. Barium Titanate... 47
   f. Alums... 50
9. Spontaneous Polarization and Spontaneous Strain.................. 51
   a. General Remarks.. 51
   b. KH_2PO_4-Type Ferroelectrics................................. 52

KÄN

KÄNZIG, Werner

c. Rochelle Salt... 55
d. Barium Titanate-Type Ferroelectrics......................... 56
e. Other Ferroelectrics...................................... 59
10. The Transition Heat...................................... 60
a. General Remarks.. 60
b. KH_2PO_4-Type Ferroelectrics................................. 60
c. Rochelle Salt... 62
d. Barium Titanate-Type Ferroelectrics......................... 62
e. Pyrochlore-Type Ferroelectrics.............................. 64
f. Other Ferroelectrics...................................... 64
11. The Influence of External Stresses on the Ferroelectric Phase Transitions... 65
a. The Influence of Electric Fields............................. 65
b. The Influence of Mechanical Forces.......................... 67
III. Thermodynamic Theory of Ferroelectric Crystals.................. 68
12. Thermodynamic Theory of Ferroelectrics That Are Piezoelectric in the Paraelectric State... 69
a. Physical Constants with Normal and with Anomalous Temperature Dependence... 69
b. Linear Relations and Corresponding Thermodynamic Potentials... 69
c. The Introduction of Saturation Terms into the Bilinear Thermodynamic Potentials... 73
d. The First-Order Piezoelectric Interaction and the Effect of "Clamping"... 77
e. The Second-Order Piezoelectric Interaction and the Effect of "Clamping"... 80
13. Thermodynamic Theory of Ferroelectrics That Are Not Piezoelectric in the Unpolarized Phase: Devonshire's Theory of $BaTiO_3$........... 81
a. Devonshire's Free Energy Functions........................... 81
b. The Cubic to Tetragonal Phase Transition..................... 83
c. The Behavior of the Cubically Clamped Crystal................ 88
IV. Optical Properties of Ferroelectrics in the Visible Range.............. 88
14. The Electro-Optic Effect.................................. 88
a. General Remarks.. 88
b. Observation of the Linear Electro-Optic Effect................. 89
c. Observation of the Second-Order Electro-Optic Effect............ 92
15. The Refractive Index..................................... 96
V. Domain Structure and Hysteresis............................. 97
16. Introduction.. 97
a. The Factors Governing the Domain Structure.................. 97
b. Experimental Evidence for Domains......................... 98
17. The Geometry of Ferroelectric Domain Structures................. 99
a. KH_2PO_4-Type Ferroelectrics................................. 99
b. Rochelle Salt... 100
c. The Domain Structure of $BaTiO_3$............................ 101
18. The Walls of Ferroelectric Domains........................... 109
a. General Remarks.. 109
b. KH_2PO_4-Type Ferroelectrics................................ 111
c. Rochelle Salt... 111
d. $BaTiO_3$.. 113

KÄNZIG, Werner

19. The Domain Structure of Minimum Free Energy.................. 114
    a. Equilibrium Domain Width in a Large Crystal Consisting of Narrow 180° Domains... 114
    b. Case in Which the Domain Width Is Comparable to the Crystal Dimensions... 117
    c. The Influence of Imperfections on the Domain Structure.......... 118
20. Dynamic Properties of Domains, Hysteresis...................... 118
    a. KH_2PO_4.. 119
    b. Rochelle Salt.. 120
    c. Barium Titanate.. 121
VI. Thermodynamic Behavior of Antiferroelectrics: Experimental.......... 124
21. Introduction.. 124
22. The Dielectric Behavior of Antiferroelectrics.................... 124
    a. $NH_4H_2PO_4$-Type Antiferroelectrics......................... 124
    b. $(NH_4)_2H_3IO_6$-Type Antiferroelectrics..................... 127
    c. Perovskite-Type Antiferroelectrics......................... 128
    d. Tungsten Trioxide, WO_3................................. 131
23. The Influence of Electric Fields on Antiferroelectric Crystals......... 131
    a. Lead Zirconate, $PbZrO_3$................................. 133
    b. Sodium Niobate, $NaNbO_3$................................ 134
    c. $NH_4H_2PO_4$-Type and Periodate-Type Antiferroelectrics........... 135
24. The Electromechanical Behavior of Antiferroelectrics.............. 135
25. The Spontaneous Strain of Antiferroelectrics.................... 135
    a. $NH_4H_2PO_4$.. 136
    b. Periodates.. 136
    c. $PbZrO_3$, $PbHfO_3$, and $NaNbO_3$....................... 136
    d. WO_3... 139
26. The Transition Heat.. 139
    a. $NH_4H_2PO_4$-Type....................................... 139
    b. Periodates.. 140
    c. Perovskite-Type Antiferroelectrics and WO_3................. 141
VII. Thermodynamic Theory of Antiferroelectrics..................... 141
27. Simple One-Dimensional Theory.............................. 141
    a. Second-Order Transition................................. 142
    b. First-Order Transition................................... 143
    c. Application to Experimental Data......................... 144
28. Three-Dimensional Theory.................................. 144
29. A Theory Which Includes the Electromechanical Behavior.......... 145
VIII. Some Optical Properties of Antiferroelectrics................... 145
30. $NH_4H_2PO_4$-Type Antiferroelectrics......................... 145
31. Perovskite-Type Antiferroelectrics........................... 146
IX. Antiferroelectric Domains.................................. 147
32. General Remarks.. 147
33. Observation of Antiferroelectric Domains...................... 147
X. The Static and Dynamic Structure of Ferroelectric and Antiferroelectric Crystals.. 148
34. The Structure of KH_2PO_4-Type Ferroelectrics................. 149
    a. Neutron and X-Ray Diffraction............................ 149
    b. Infrared and Raman Spectra.............................. 153
    c. Nuclear Magnetic Resonance............................. 154

KÄN

KÄNZIG, Werner

35. The Structure of Antiferroelectric $NH_4H_2PO_4$ 155
   a. X-Ray Diffraction 155
   b. Raman and Infrared Spectra 157
36. The Structure of the Antiferroelectric Periodates 157
37. The Structure of the Ferroelectric Tartrates 159
   a. X-Ray and Neutron Diffraction Data 159
   b. Raman Effect 161
38. The Structure of Perovskite-Type Ferroelectrics 161
   a. General Remarks 161
   b. The Structure of $BaTiO_3$ in the Tetragonal Phase 162
   c. The Structure of $PbTiO_3$ in the Tetragonal Phase 164
   d. The Thermal Oscillation Amplitudes 165
   e. Infrared and Optical Absorption of $BaTiO_3$ and $SrTiO_3$ 166
   f. Nuclear Resonance 166
   g. Paramagnetic Resonance 167
39. The Structure of Perovskite-Type Antiferroelectrics 167
   a. The Structure of $PbZrO_3$ in the Antipolarized Phase 167
   b. The Structure of $NaNbO_3$ in the Orthorhombic Phase at Room
      Temperature 168
   c. Other Interesting Perovskite-Type Dielectrics 169
40. The Structure of WO_3 170
41. The Structure of Pyrochlore-Type Ferroelectrics ($Cd_2Nb_2O_7$) 171
42. The Structure of Ferroelectric Lead Metaniobate, $PbNb_2O_6$ 172
43. Ferroelectrics with the Pseudo-Ilmenite Structure ($LiTaO_3$, $LiNbO_3$) .. 172
44. The Structure of the Alums 172
XI. Isotope Effects 173
45. Deuterated Rochelle Salt 173
46. Deuterated KH_2PO_4-Type Ferroelectrics 173
47. Deuterated $NH_4H_2PO_4$-Type Antiferroelectrics 174
48. Deuterated $(NH_4)_2H_3IO_6$ and $Ag_2H_3IO_6$ 175
XII. Ferroelectricity and Antiferroelectricity of Solid Solutions 175
49. Solid Solutions with the Tetragonal KH_2PO_4 Structure 175
50. Solid Solutions with the Rochelle Salt Lattice 176
51. Solid Solutions between Perovskite-Type Ferroelectrics, Antiferro-
    electrics, and Paraelectrics 176
   a. Solid Solutions between a Ferroelectric and an Antiferroelectric 177
   b. Solid Solutions between a Paraelectric and an Antiferroelectric 178
   c. Solid Solutions between a Paraelectric and a Ferroelectric 178
   d. Solid Solutions between Two Ferroelectrics 179
   e. Solid Solutions between Two Paraelectrics 180
52. Miscellaneous Mixed Crystals 180
XIII. Molecular Models 180
53. Electronic Polarization 180
   a. Sublattices, Internal Fields, and Electronic Polarizabilities 180
   b. The Conditions for an Electronic Polarization Catastrophe 182
54. Anomalous Electronic Polarizabilities 184
55. Ionic Polarization (Lattice Polarization) 187
   a. Static Considerations of the Interaction of the Ionic and Electronic
      Polarization 187
   b. Ferroelectricity *versus* Antiferroelectricity 188

KÄNZIG, Werner

56. Ionic Polarization and Lattice Vibrations.......................... 189
57. Ferroelectricity as an Order-Disorder Problem...................... 191
    a. Permanent Pseudodipoles in an Internal Field Described by Lorentz Factors... 192
    b. Permanent Pseudodipoles in an Onsager Internal Field........... 193
    c. The Slater Model for KH_2PO_4................................ 194
    d. The Spherical Model... 196

KELLER, Joachim: *see* FULDE, Peter

KELLER, P.   SUPPLEMENT 14, 1978

Liquid-Crystal Synthesis for Physicists

P. KELLER AND L. LIEBERT

Laboratoire de Physique des Solides, Université Paris-Sud. Orsay, France

I. Introduction 20
II. A General Presentation 22
III. Schiff's Bases 23
    1. Preparation of *p*-Alkyloxybenzylidene-*p'*-Alkylanilines 24
    2. Preparation of *p*-Substituted Benzylidene-*p'*-*n*-Alkyloxyaniline 36
    3. Preparation of *p*-Substituted Benzylidene-*p'*-Aminocinnamate 41
    4. Preparation of Di Schiff's Bases 46
    5. Conclusion 48
IV. *p-p'*-Disubstituted Azoxybenzene 48
    6. Duplicative Oxidation 49
    7. Duplicative Reduction Method 51
    8. Conclusion 54
V. Esters 54
    9. Preparation of *p-p'*-Disubstituted Phenylbenzoates 55
    10. Preparation of Phenyl 4-Benzoyloxybenzoate Derivatives 67
    11. Conclusion 69
VI. Purification 70
    12. Crystallizations 70
    13. Extraction Technique 71
    14. Distillation 73
    15. Conclusion 74
    16. Melting Points 74
    17. Other Purity Determination Methods 75
VII. Conclusion 75

KEL

KELLY, M.Y.　　　　VOLUME 35, 1980

Applications of the Recursion Method to the Electronic Structure from an Atomic Point of View

M.J. KELLY

Cavendish Laboratory, University of Cambridge, Cambridge, England

I. Introduction 296
II. The Recursion Method in Practice.. 299
   1. The Tight-Binding Scheme and Orthonormal Orbitals 299
   2. The Local Density of States... 301
   3. Bond Orders.. 302
   4. Total Energies 303
   5. Methods for Calculating the Local Density of States.................... 305
   6. The Recursion Method... 307
   7. The Starting Orbital for the Recursion Method 311
   8. Cluster Calculations, Continued-Fraction Processing, and Total Energy
      Calculations 313
   9. Self-Consistency.. 316
  10. Similar Schemes... 317
  11. Summary: When to Use the Recursion Method........................ 319
III. Surface Electronic Structure 320
  12. Introduction.. 320
  13. The Electronic Structure of Ideal Solid Surfaces....................... 321
  14. The Electronic Structure of Nonideal Surfaces 328
  15. The Electronic Structure of Ordered Overlayers 330
  16. Aspects of Chemisorption Theory..................................... 332
  17. Interactions between Adsorbed Species 336
IV. The Electronic Structure of Random Networks........................... 339
  18. Introduction.. 339
  19. Fourfold Coordinated Networks....................................... 340
  20. Threefold Coordinated Networks 347
  21. Other Networks, Including Metallic Glasses 353
V. Phonon Spectra.. 356
  22. Phonons in the Harmonic Approximation............................. 356
  23. Surface and Interface Phonons 357
  24. Phonons on Random Networks 358

KELLY, M.J.

VI. Problems Involving Small Energy Differences 363
    25. Introduction 363
    26. The Laves and Rare-Earth Crystal Structures......................... 364
    27. Frozen Phonons... 366
    28. Stoner Magnetic Energies and Their Contribution to Crystal Structure
        Stability... 368
    29. Susceptibilities 370
VII. Miscellany 372
    30. Configuration Averaging in Disordered Systems 372
    31. A Recursion Method Perturbation Theory with an Application to Anderson
        Localization 375
    32. Ultraviolet Photoemission 377
    33. Fano Effects in the Electronic Structure of d Band Metals............... 378
    34. Spin Waves in Heisenberg Systems.................................. 378
VIII. Concluding Remarks 380
    35. The Electronic Structure of Molecules: A Future Application 380
    36. Factors Determining Magnetic Moments in Transition Metal Systems:
        A Current Application 381
    37. Summary.. 382
    Note Added in Proof 382

KELTON, K.F.

VOLUME 45, 1991

Crystal Nucleation in Liquids and Glasses

K.F. KELTON

Department of Physics
Washington University
St. Louis, Missouri

I. Introduction 75
    1. Motivation... 75
    2. Background 78
II. Classical Theory: Steady-State Nucleation 83
    3. Basic Theory 83
    4. Cluster Translation and Rotation 91

KEL–KER

KELTON, K.F.

III.	Classical Theory: Time-Dependent Nucleation		93
	5.	Phenomena	93
	6.	Analytical and Closed-Form Solutions	96
	7.	Numerical Solutions	99
	8.	Preexisting Cluster Distributions	102
IV.	Experimental Studies of Nucleation in Undercooled Liquids		103
	9.	Undercooling of Polymorphically Crystallizing Liquids	103
	10.	Nucleation Rate Measurements	117
V.	Experimental Studies of Nucleation in Glasses		123
	11.	Experimental Methods and Analysis	123
	12.	Crystallization of Metallic Glasses	127
	13.	Silicate Glasses	135
	14.	A Test of the Kinetic Model of Classical Nucleation Theory	151
VI.	Computer Simulations of Steady-State Nucleation		155
	15.	Monte Carlo Calculations	155
	16.	Molecular Dynamics	157
VII.	Beyond Classical Nucleation Theory		159
	17.	Analysis of the Classical Theory	159
	18.	Curvature and Anisotropy Effects on σ	161
	19.	Field-Theoretic Models of Steady-State Nucleation	162
	20.	Further Treatments of Non–Steady-State Nucleation	169
VIII.	Concluding Remarks		172
		Appendix A: Approximations to ΔG	173
		Acknowledgments	177

KERN, R.: *see* BUSCH, G.A.

KEYES, Robert W.   VOLUME 11, 1960

The Effects of Elastic Deformation on the Electrical Conductivity of Semiconductors

ROBERT W. KEYES

Research Laboratories, Westinghouse Electric Corporation, Pittsburgh, Pennsylvania

I. Introduction 149
II. Phenomenological Description of Resistance and Piezoresistance 150
   1. Notation 151
   2. The Requirements of Crystal Symmetry 152
   3. Piezogalvanomagnetic Effects 153
III. Measurement of Piezoresistance 156
   4. Effects of Hydrostatic Pressure 156
   5. Effects of Shear Stress 157
   6. Dimensional Corrections 160
   7. Adiabatic and Isothermal Constants 161
   8. Transformations of the Piezoresistance Tensor 161
IV. Properties of Semiconductors 163
   9. Energy Level Structure 163
   10. Deformation Potential Model of the Effect of Strain 167
   11. Transport Properties 168
V. The Effects of Hydrostatic Pressure 169
   12. Effects of Pressure on the Band Structure 170
   13. The Thermodynamic Interpretation 179
   14. The Bonding Interpretation 184
VI. The Effects of Shear Strain 187
   15. The Multivalley Semiconductor 187
   16. Semimetals 210
   17. Degenerate Bands 211
VII. Related Phenomena 213
   18. The "Minor Effects" 213
   19. Piezo-Optical Experiments 214
   20. Dependence of Energy Gap on Chemical Composition 218

KEY

KEYES, Robert W.   VOLUME 20, 1967

Electronic Effects in the Elastic Properties of Semiconductors

ROBERT W. KEYES

IBM Thomas J. Watson Research Center, Yorktown Heights, New York

I. Introduction.. 37
II. Phenomenological Description of Elastic Constants....................... 39
III. Effect of Free Electrons on Elastic Properties........................... 40
    1. Ordinary Elastic Constants of Multiband Semiconductors.............. 40
    2. Crystal Stability.. 52
    3. Electronic Relaxation and Attenuation of Elastic Waves............... 53
    4. Elastic Constants of Other Semiconductors........................... 56
    5. Thermal Properties.. 59
    6. Third-Order Elastic Constants....................................... 63
IV. Electronic Strain.. 67
    7. Photostriction.. 68
    8. Doping by Electron Irradiation...................................... 68
    9. Electronic Magnetostriction in Bismuth.............................. 70
    10. Theory of Magnetostriction and Magnetoelasticity in Germanium and
        Silicon... 74
V. Electrons on Donors.. 78
    11. Effect of Donors on Ordinary Elastic Constants..................... 78
    12. Effect of Donors on Thermal Resistance............................. 83
VI. Conclusion... 89

KIMERLING, L.C.   VOLUME 50, 1996

Light Emission from Silicon

L.C. KIMERLING, K.D. KOLENBRANDER, J. MICHEL, J. PALM, AND L. GIOVANE

Department of Materials Science and Engineering Massachusetts Institute of Technology, Cambridge, MA

I.	Overview	334
	1. Introduction	334
	2. Black Body Radiation	334
	3. Luminescence	335
	4. Direct and Indirect Gap Materials	336
	5. Recombination Processes	337
II.	Intrinsic Luminescence in Silicon	339
III.	Extrinsic Luminescence	342
	6. Bound Excitons	344
	7. Isoelectronic Bound Exciton Centers	347
	8. Dislocation Luminescence	349
	9. Deep-State Luminescence	352
	10. Rare-Earth Centers	353
IV.	Silicon Alloys	360
V.	Strain Fields	363
	11. Band Structure	363
	12. Coherent Heterostructures	365
VI.	Nanostructured Silicon	366
	13. Quantum Confinement	367
	14. Nanostructured Silicon Materials	369
	15. Surface Effects	370
	16. Structure/Size Issues	371
VII.	Applications	373
	17. Materials Characterization	373
	18. Hot Electron Effects	376
	19. Light-Emitting Diodes	377
	20. Nanostructured Silicon	379
VIII.	Conclusions	380

KIT

KITTEL, C.   VOLUME 3, 1956

Ferromagnetic Domain Theory

C. Kittel

Department of Physics, University of California, Berkeley, California

AND

J.K. Galt

Bell Telephone Laboratories, Inc., Murray Hill, New Jersey

I. Survey of Domain Theory... 439
   1. Introduction... 439
   2. The Domain Assumption... 440
   3. The Origin of Domains... 445
      a. Size of Domains.. 450
      b. Polycrystalline Materials.................................. 450
   4. Coercive Force, Hysteresis, and Reversible Permeability............ 452
      a. Reversible Permeability.................................... 456
      b. Barkhausen Effect.. 456
 II. Contributions to the Domain Energy.............................. 457
   5. Exchange Energy... 457
      a. Exchange Energy on the Atomic (Heitler-London-Heisenberg) Model 459
   6. Anisotropy Energy... 463
      a. Mathematical Representation of the Anisotropy Energy........... 463
      b. Physical Origin of the Anisotropy Energy...................... 464
   7. Magnetoelastic Energy... 467
      a. The Minimal Equations...................................... 469
      b. Relation to Magnetostriction Constants..................... 470
      c. Isotropic Magnetostriction................................. 471
   8. Magnetostatic Energy.. 472
      a. Interaction of Permanent Magnet with External Field........... 472
      b. Self-Energy of a Permanent Magnet.......................... 473
      c. Ellipsoidal Specimen....................................... 473
III. The Bloch Wall.. 473
   9. Introductory Remarks.. 473
  10. Estimate of Thickness and Energy of Bloch Wall................. 475
  11. 180 Degree Walls in (100) Plane of Iron....................... 476
      a. Value of the Wall Energy................................... 478
      b. Wall Thickness... 479
 IV. Theoretical Domain Structures................................... 481
  12. Introduction.. 481
  13. Flux Closure Domain Configurations............................ 483

KITTEL, C.

a. Cubic Crystals with Positive Anisotropy Constant 486
b. General Case 487
V. Experimental Studies of Domains 490
14. Magnetic Powder Pattern Technique 491
    a. Experimental Powder Pattern Technique 491
    b. Theory of the Magnetic Powder Pattern Technique 493
15. Kerr Magneto-Optic Effect Technique 495
16. Selection of Results of Powder Pattern Work 498
    a. Scratch Technique 498
    b. Correlation between Magnetization Changes and Boundary Displacement 499
    c. Domain Structure around Cavities 501
    d. Domain Structure at Grain Boundaries 502
VI. Magnetic Properties of Small Particles 502
17. Critical Particle Size for Single Domain Behavior 504
    a. Low Anisotropy 505
    b. High Anisotropy, Cubic Crystal 506
    c. High Anisotropy, Uniaxial Crystal 506
18. Coercive Force of Small Particles 507
    a. Coercive Force Resulting from Magnetocrystalline Anisotropy 507
    b. Coercive Force Resulting from Anisotropy of Particle Shape 509
    c. Coercive Force Resulting from Longitudinal Stress 511
    d. Dependence of the Coercive Force on Particle Size 511
    e. Dependence of the Coercive Force on the Density of Packing 512
    f. Rock Magnetism 513
19. Fine Particles in Permanent Magnets: Alnico 5 513
    a. Origin of the Coercive Force 514
    b. Source of Particle Anisotropy 515
    c. Heat Treatment in a Magnetic Field 517
VII. Coercive Force of Bulk Material 519
20. General Remarks 519
21. Kersten Inclusion Theory 521
    a. Initial Permeability on the Inclusion Model 523
22. Strain Theory 524
23. Néel Theory: Demagnetizing Fields and the Effects of Thermal Fluctuations 525
24. Goodenough Lamellar Precipitate and Grain Boundary Theory 528
25. Experimental Observations 529
VIII. Motion of Domain Walls 534
26. Equation of Motion: Small Amplitude Motion of 180 Degree Walls ... 534
27. Equation of Motion: Large Amplitude Motion of 180 Degree Walls ... 535
28. Experimental Measurements on Domain Wall Motion 535
29. Domain Wall Stiffness 545
30. Domain Wall Mass 546
31. Viscous Damping Parameter 548
32. The 90 Degree Wall 551
33. Theory of the Viscous Damping Parameter β 551
34. Numerical Results and Comparisons of Experiment and Theory 555
Appendix A. Mathematical Expressions for the Anisotropy of Cubic Crystals 557

KIT–KLE

KITTEL, C. VOLUME 22, 1968

Indirect Exchange Interactions in Metals

C. KITTEL

Department of Physics, University of California, Berkeley, California

I. Introduction...	1
II. Wave-Vector Dependent Susceptibility................................	4
III. Spectrum of the Effective Magnetic Field.............................	7
1. Single Moment...	7
2. Ordered Spin Array..	7
IV. Magnetization at Large Distances.....................................	8
V. Spin Susceptibility of a Free Electron Gas............................	10
VI. Derivation of $F_3(q)$ in Three Dimensions...........................	13
VII. Derivation of $F_2(q)$ in Two Dimensions............................	14
VIII. Discussion of Range Function in Three Dimensions.....................	15
IX. Correlation and Exchange Effects on χ_q.........................	18
X. Effect of Impurities on the Range Function...........................	20
XI. Nonspherical Fermi Surfaces...	21
XII. Divergences in Higher Order...	22
XIII. Experimental Evidence..	22
Appendix A: Range Function in One Dimension........................	22
Appendix B: Range Function in Three Dimensions.....................	24
Appendix C: Range Function for Two Valleys........................	25

KLEMENS, P.G. VOLUME 7, 1958

Thermal Conductivity and Lattice Vibrational Modes

P.G. KLEMENS

Division of Physics, National Standards Laboratory, C.S.I.R.O., Sydney, Australia

I. Introduction...	1
1. Introduction..	1
II. Theory of Heat Transport by Lattice Vibrations.......................	7
2. Normal Modes of a Lattice (Résumé)..............................	7
3. Quantized Lattice Waves...	11
4. Formal Theory of Thermal Conductivity...........................	13
5. The Interaction Processes..	15
a. Scattering by Static Imperfections...........................	15
b. Interactions due to Anharmonicities..........................	24
c. Boundary Scattering..	28
6. Approximate Solution of the Transport Equation...................	29
7. Boundary Resistance...	41

KLEMENS, P.G.

III. Thermal Conductivity of Nonmetals.................................. 45
   8. Thermal Conductivity at High Temperatures........................ 45
   9. Intrinsic Thermal Resistance at Low Temperatures.................. 50
  10. The Size Effect.. 55
  11. Thermal Resistance due to Imperfections.......................... 59
  12. Polycrystalline and Highly Disordered Materials................... 64
     a. Polycrystalline Solids....................................... 65
     b. Glasses.. 67
     c. Effects of Radiation Damage.................................. 69
IV. Thermal Conductivity of Metals, Alloys, and Semiconductors........... 70
  13. Electronic Thermal Conductivity (Résumé)......................... 70
  14. Scattering of Lattice Waves by Conduction Electrons.............. 74
  15. Lattice Component of Thermal Conductivity........................ 78
  16. Separation of Observed Thermal Conductivity into Electronic and
     Lattice Components.. 82
  17. The Intrinsic Lattice Thermal Conductivity (Observations)........ 84
  18. The Effect of Crystal Imperfections.............................. 90
  19. Semiconductors: Resistance due to Free Electrons................. 94

KLICK, Clifford C.   VOLUME 5, 1957

Luminescence in Solids

CLIFFORD C. KLICK AND JAMES H. SCHULMAN

U.S. Naval Research Laboratory, Washington, D.C.

  I. Introduction.. 97
 II. Luminescence Involving Absorption and Emission Processes in Simple
    Centers... 99
   1. Classical Configuration Coordinate Curve Model................... 100
   2. Quantum-Mechanical Modifications of Configuration Coordinate Curves 104
   3. Calculation of Configuration Coordinate Curves from Experiment... 109
   4. Theoretical Calculation of Configuration Coordinate Curves....... 113
   5. Continuous Dielectric Model..................................... 116
   6. Summary of Present Status....................................... 118
 III. Luminescence Involving Energy Transfer with No Movement of Charge... 119
   7. Energy Transfer by Emission and Reabsorption of Photons.......... 119
   8. Energy Transfer by Nonradiative Processes........................ 121
   9. Impurity-Sensitized Luminescence................................ 122
     a. General Description of Phenomena............................. 122
     b. Interpretations of Impurity Sensitization................... 126
     c. Theoretical Treatment of Resonance Transfer................. 131
     d. Host-Sensitized Luminescence................................ 139
  10. Concentration Quenching and Quenching in Pure Crystals........... 142
  11. Summary of Present Status.. 144
 IV. Luminescence in Systems Involving Transfer of Charge................. 144
  12. Preparation and Properties of ZnS-Type Phosphors................. 146
  13. Transport and Recombination Phenomena............................ 154
  14. Electroluminescence.. 166
  15. Summary of Present Status.. 171

KNI

KNIGHT, W.D. VOLUME 2, 1956

Electron Paramagnetism and Nuclear Magnetic Resonance in Metals

W.D. KNIGHT

University of California, Berkeley, California

I. Introduction... 93
   1. Background.. 93
   2. Simple Theory of the Line Shift in Metals......................... 94
   3. Outline of Problems... 95
II. Theory.. 97
   4. Calculation of the Line Shift................................. 97
   5. The Electron Spin Paramagnetism............................. 98
   6. The Electron Wave Functions................................ 101
   7. The ξ Factor... 103
   8. The Effects of Temperature, Volume, and Magnetic Field........... 105
   9. The Effects of Impurities................................... 106
   10. The Chemical Shifts....................................... 108
III. Experiments.. 110
   11. The Apparatus... 110
   12. The Samples... 111
   13. The Measurements.. 112
IV. Results.. 113
   14. The Magnetic Susceptibility................................ 113
   15. The Hyperfine Coupling Constant............................ 119
   16. The Resonance Line Shifts.................................. 121
   17. Comparison of Values for χ_p............................. 124
   18. The ξ Factor... 125
   19. The Effects of Temperature and Magnetic Field.................. 126
   20. The Line Shift in Alloys.................................... 129
   21. Application to the Problem of Superconductivity.................. 133
V. Summary.. 134

KNIGHT, W.D.: *see* DE HEER, Walt A.

KNOX, Robert S. VOLUME 4, 1957

Bibliography of Atomic Wave Functions

ROBERT S. KNOX

Institute of Optics, University of Rochester, Rochester, New York

I. Introduction.. 413
II. Alphabetical List of Atoms and Ions.............................. 416

KNOX, Robert S. SUPPLEMENT 5, 1963

Theory of Excitons

ROBERT S. KNOX

Department of Physics and Astronomy, University of Rochester, Rochester, New York

Preface   v

I. Introduction   1
   1. Survey   1

II. The Electronic Structure   7
   2. General Case: Use of the One-Electron Approximation . . .   7
   3. Frenkel's Case: Tight Binding 21
   4. Wannier's Case: Weak Binding 37
   5. Intermediate Cases 59
   6. Effects of Static External Fields 74
   7. Special Topics. 87

III. Absorption and Dispersion of Light by Nonmetallic Solids . . . 103
   8. Classical Theory of the Optical Effects 103
   9. Semiclassical Theory of Optical Absorption 112
   10. Dynamical Effects of Phonons 136
   11. Anomalous Waves and Spatial Dispersion 163

IV. Transport Phenomena and Related Topics 169
   12. Theory of Exciton Transport Phenomena 169
   13. Radiative Decay 188

V. Summary 191

AUTHOR INDEX 195

SUBJECT INDEX 201

KOE–KOH

KOEHLER, J.S.L: *see* SEITZ, Frederick

KOHN, W. VOLUME 5, 1957

Shallow Impurity States in Silicon and Germanium

W. KOHN

Carnegie Institute of Technology, Pittsburgh, Pennsylvania

I. Introduction... 258
II. Empirical Properties.. 261
   1. Energy Levels... 261
      a. Ionization Energies................................... 261
      b. Spectra of Excited States............................. 264
   2. Spin Resonance.. 266
      a. Electron Spin Resonance............................... 266
      b. Double Resonance...................................... 266
   3. Static Magnetic Susceptibility 271
III. Structure of Donor States................................. 271
   4. Conduction Bands of Silicon and Germanium................ 271
      a. Silicon... 272
      b. Germanium... 273
   5. Effective Mass Theory of Donor States.................... 274
      a. Single Band Minimum at $\mathbf{k} = 0$............... 276
      b. Several Conduction Band Minima........................ 281
      c. Matrix Elements for Radiative Transitions............. 283
   6. Numerical Results and Comparison with Experiments........ 285
      a. Energy Levels... 285
      b. Wave Functions.. 287
   7. Corrections to the Effective Mass Formalism............. 289
      a. General Considerations................................ 289
      b. Corrected Wave Functions.............................. 290
      c. Comparison with Experiment............................ 294
IV. Structure of Acceptor States............................... 297
   8. Valence Bands of Silicon and Germanium.................. 297
      a. Silicon... 297
      b. Germanium... 300
   9. Effective Mass Equations for Acceptor States............ 300
   10. Approximate Solutions and Comparison with Experiment... 301
      a. Germanium... 301
      b. Silicon... 303
V. Effects of Strains and of Static Electric and Magnetic Fields............. 306
   11. Strains... 306
      a. Donor States.. 307
      b. Acceptor States....................................... 309
   12. Stark Effect.. 311

KOHN, W.

13. Static Magnetic Susceptibility.................................... 312
    a. Donor States.. 312
    b. Acceptor States... 314
14. Magnetic Field Dependence of the Ionization Energy................ 315
VI. Interaction with Lattice Vibrations............................... 316
15. Absorption Line Widths... 316
    a. Optical Absorption Line Widths............................... 317
    b. Absence of Paramagnetic Resonance in Acceptor States......... 319
Appendix: Deep-Lying Impurity Levels................................. 319

KOLENBRANDER, K.D.: *see* KIMERLING, L.C.

KONDO, J.   VOLUME 23, 1969

Theory of Dilute Magnetic Alloys

J. Kondo

Electrotechnical Laboratory, Tokyo, Japan

I. The s-d ExchangeModel.. 184
   1. Introduction.. 184
   2. Derivation of the s-d Exchange Hamiltonian................ 185
   3. Fröhlich–Nabarro–Zener Interaction........................... 188
   4. Ruderman–Kittel–Kasuya–Yosida Interaction.................... 188
   5. Korringa Relaxation... 191
II. Formation of Local Moments....................................... 193
   6. One-Dimensional Models.. 194
   7. Friedel's Sum Rule.. 196
   8. k-Independent Impurity Potential............................ 199
   9. Anderson's Mixing Model....................................... 202
   10. Spin-Splitting of the Localized Level........................ 204
   11. Spin-Splitting on Anderson's Model........................... 205
   12. Derivation of the s-d Hamiltonian from Anderson's Model.. 210
III. Logarithmic Divergence in Higher Approximations................. 213
   13. Resistivity and Magnetoresistance in the First Born Approximation..... 213
   14. Theory of Resistance Minimum................................. 217
   15. Higher-Order Terms in Other Physical Parameters.............. 223

KONDO, J.

IV. Removal of the Divergence at Low Temperatures........................ 231
   16. Abrikosov's Theory... 231
   17. Suhl's Theory... 234
   18. Nagaoka–Hamann–Bloomfield Theory............................... 241
   19. Physical Parameters at Low Temperatures........................ 247
   20. Consideration of the Ground-State Energy....................... 252
   21. Free Energy in Suhl's Theory.................................. 256
   22. Effects of Ordinary Scattering on Exchange Scattering......... 261
   23. Other Approaches to the Problem................................ 263
   24. Exchange Scattering in Superconductors........................ 275
   Appendix A: Evaluation of Eq. (11.5)............................... 279
   Appendix B: Proof of Eq. (18.38).................................. 279
   Appendix C: Evaluation of the First Two Terms of Eq. (21.5)........ 280

KOSTER, G.F. VOLUME 5, 1957

Space Groups and Their Representations

G.F. Koster

Massachusetts Institute of Technology, Cambridge, Massachusetts

  I. Nature and Properties of the Space Groups........................... 174
    1. Orthogonal Transformations................................... 175
    2. Space Groups... 177
    3. Point Groups... 180
    4. Bravais Lattices... 194
      a. Cubic System... 196
      b. Tetragonal System...................................... 199
      c. Orthorhombic System.................................... 202
      d. Monoclinic System...................................... 205
      e. Triclinic System....................................... 207
      f. Trigonal and Hexagonal Systems......................... 208
 II. Irreducible Representations of Space Groups........................ 211
    1. Introduction... 211
    2. General Theory... 211
    3. Simplifications.. 222
III. Examples... 224
    1. Simple Cubic O_h^1... 225
      a. General Point.. 226
      b. General Points in a Symmetry Plane..................... 226
    2. Body-Centered Cubic O_h^9................................. 227
      a. General Point on the Surface........................... 227
    3. Face-Centered Cubic O_h^5................................. 228
      a. General Point on the Surface........................... 228
    4. Zincblende Structure T_d^2................................ 229
      a. General Point.. 229
      b. General Point on a Symmetry Plane...................... 229
    5. Diamond Structure O_h^7................................... 230
      a. General Point.. 231
      b. General Points on Symmetry Planes...................... 231
    6. Hexagonal Close-Packed D_{6h}^4.......................... 234
      a. General Point.. 234

KOSTER, G.F.

IV. Double Groups... 238
    1. General.. 238
V. Time Reversal.. 249
VI. History... 252

KOTHARI, L.S.   VOLUME 8, 1959

Interaction of Thermal Neutrons with Solids

L.S. KOTHARI AND K.S. SINGWI

Atomic Energy Establishment, Trombay, Bombay, India

  I. Introduction.. 110
 II. General Theory of Neutron Scattering................................. 112
     1. Pseudo-Potential... 112
     2. Decomposition of the Displacement Vector......................... 115
     3. Elastic Scattering Cross Section................................. 116
        a. Incoherent Cross Section...................................... 121
        b. Coherent Cross Section.. 121
     4. Inelastic Scattering Cross Section............................... 124
        a. Incoherent Cross Section...................................... 126
        b. Coherent Cross Section.. 128
     5. Limiting Cases... 131
        a. One-Phonon Cross Section...................................... 131
        b. Multiphonon Cross Section..................................... 132
        c. Placzek Expansion... 134
     6. Einstein Approximation... 137
III. Cold Neutron Scattering... 139
     7. Scattering from Isotropic Lattices.............................. 139
        a. Magnesium... 140
        b. Aluminum.. 141
     8. Scattering from Graphite... 144
     9. Scattering in Lead near Melting Point............................ 149
 IV. Crystal Dynamics and Inelastic Scattering of Neutrons............... 151
     10. Dispersion Relation... 152
         a. Study with Infinite Wavelength Neutrons...................... 154
         b. Study with Finite Wavelength Neutrons........................ 156
         c. Optical Modes.. 160
         d. Excitations of Liquid Helium II.............................. 162
     11. Frequency Distribution Function................................. 164
  V. Slowing Down of Neutrons near Thermal Equilibrium................... 168
     12. Gas Model... 168
     13. Cross Sections on Debye Model................................... 171
         a. Scattering Cross Section..................................... 172
         b. Transport Cross Section...................................... 174
     14. Slowing-Down Parameters... 180
         a. Mean Logarithmic Energy Decrement ξ and Mean Energy Loss per
            Collision $<E_1 - E_2>$ 180
         b. Age... 183
         c. Time Behavior... 184

KOT–KRÖ

KOTHARI, L.S.

VI. Appendix.. 186
    15. Multiphonon Cross Section.. 186
    16. Lattice Dynamics.. 187
    17. Slowing Down of Neutrons... 188

KRÖGER, F.A.   VOLUME 3, 1956

Relations between the Concentrations of Imperfections in Crystalline Solids

F.A. KRÖGER AND H.J. VINK

Philips Research Laboratories, N.V. Philips' Gloeilampenfabrieken, Eindhoven-Netherlands

I. Pure Stoichiometric Compounds....................................... 310
    1. Atomic Disorder... 310
    2. Electronic Disorder... 315
      a. Electronic Disorder Not Involving Atomic Imperfections........ 315
      b. Electronic Disorder Involving Atomic Imperfections........... 318
    3. Complete Equilibrium Involving Both Electronic and Atomic Disorder. 323
      a. Equilibrium at High Temperatures............................ 324
      b. Processes Taking Place during Cooling........................ 325
    4. Migration of Atomic Imperfections; Ionic Conductivity.............. 326
    5. Electronic Conductivity Arising from Lattice Defects............... 328
II. Nonstoichiometric Compounds... 329
    6. Introduction... 329
    7. Reactions Involving Crystal and Vapor........................... 329
    8. Complete Equilibrium Crystal-Vapor for a Simple Case with Frenkel
      Disorder.. 333
    9. A Method for Obtaining an Approximate Solution of Multi-Equation
      Relations Such as Those of Section 8............................ 335
    10. Complete Equilibrium between Crystal and Vapor for a Crystal Contain-
      ing Completely Ionized Frenkel Defects.......................... 342
    11. A More Complicated Case of Schottky-Wagner Disorder............. 344
    12. Equilibrium between Crystal and Vapor for a Crystal MX with Anti-
      structure Disorder... 347
      a. Pure Antistructure Disorder................................. 347
      b. Combination of Antistructure and Schottky-Wagner Disorder...... 349
    13. Cooling; the State of the Crystals at Low Temperatures............. 352
    14. The Deviation from the Simple Stoichiometric Ratio as a Function of
      the Atmosphere.. 357
    15. The Position of the Fermi Level................................. 360
    16. Practical Limitations... 362
    17. Comparison with Experiment..................................... 363
    18. The Effect of Heating in a Stream of Inert Gas................... 369
    19. Equilibria Vapor-Liquid and Solid-Liquid........................ 370
    20. Surface Layers... 372
III. Crystals Containing Foreign Atoms................................... 373
    21. Introduction.. 373

KRÖGER, F.A.

22. The Sites Occupied by Foreign Atoms 374
23. Energy Levels Due to Foreign Atoms 375
24. General Remarks Concerning Equilibria Solid-Vapor for Compounds
    Containing Foreign Atoms 378
25. Equilibrium between Crystal and Vapor for Binary Compounds *MX*
    Containing Foreign Atoms with Deviating Valence at Normal Lattice
    Sites 379
    a. One Level per Imperfection 379
    b. Several Levels per Imperfection 388
    c. Crystals Containing One Type of Foreign Atoms in Various Positions 390
    d. Practical Limitations 391
26. The Influence of Foreign Atoms on the Properties of Crystals 392
    a. Ionic Conductivity, Diffusion, and Dielectric Loss 392
    b. Electronic Conductivity 396
    c. Optical Properties 404
    d. Magnetic Properties 405
    e. Deviations from the Simple Stoichiometric Ratio 406
27. Comparison of Compounds and Monatomic Crystals 407
28. Crystals Containing Two Types of Foreign Atoms 408
29. Controlling the Charge of Foreign Atoms 409
IV. Solubility Relations 416
30. Introduction 416
31. Equilibrium between Crystal and Vapor 418
    a. CdS Containing Chlorine Alone 419
    b. CdS Containing Silver Alone 422
    c. CdS Containing Chlorine as Well as Silver 422
    d. Practical Limitations; Solubility in Contact with a Compound of the
       Foreign Elements 425
32. Comparison with Experiment 427
33. Equilibrium between Solid and Liquid 428
V. Comparison of Various Models Used in Describing a Crystal 431
    34. A Justification of the Atomic Notation; The Problem of Determining the
       Equilibrium Constants 431
35. Comparison of the Atomic and Ionic Notation 432

KUB

KUBO, Ryogo   VOLUME 17, 1965

Quantum Theory of Galvanomagnetic Effect at Extremely Strong Magnetic Fields

RYOGO KUBO AND SATORU J. MIYAKE

Department of Physics, The University of Tokyo, Tokyo, Japan

AND

NATSUKI HASHITSUME

Department of Physics, Ochanomizu University, Tokyo, Japan

I. Introduction.. 270

II. Basic Theory... 274
   1. Motion of Crystal Electrons in a Magnetic Field..................... 274
   2. General Expression of the Conductivity Tensor...................... 280
   3. Formulas for the Case of Elastic Scattering........................ 286
   4. Asymptotic Behavior of Conductivity at Extremely Strong Fields....... 289
   5. Logarithmic Divergence in the Perturbational Treatment.............. 295

III. Inelastic Collision with Phonons...................................... 299
   6. Effect of Inelasticity on the Logarithmic Divergence................. 299
   7. Formula for Scattering by Phonons.................................. 302
   8. Acoustic Phonons.. 307
   9. Optical Phonons and Piezoelectric Interactions...................... 316

IV. Collision Broadening.. 317
   10. Effect of Collision Broadening on the Logarithmic Divergence......... 317
   11. Damping Theoretical Formulation................................... 320
   12. Davydov–Pomeranchuk Theory...................................... 322
   13. Scattering by Gaussian Potential: Short-Range Force................. 326
   14. Collision Broadening versus Inelasticity............................ 333

V. Non-Born Scattering... 336
   15. Effect of Interference of Waves on the Logarithmic Divergence........ 336
   16. Conductivity Tensor Expressed in Terms of Scattering Operators....... 337
   17. Skobov–Bychkov Theory: Short-Range Force......................... 340
   18. Method of Partial Waves... 343
   19. Non-Born Scattering versus Collision Broadening.................... 353

   Appendix A. Collision Broadening Effect upon Oscillatory Behavior......... 356

   Appendix B. Approximate Form of the Green Function in a Magnetic Field.. 362

KWOK, Philip C.K.   VOLUME 20, 1967

Green's Function Method in Lattice Dynamics

PHILIP C.K. KWOK

IBM Thomas J. Watson Research Center, Yorktown Heights, New York

Introduction 214
I. Dynamics of Lattice Vibrations 214
   1. The Hamiltonian of Lattice Vibrations. 214
   2. Expansion in Nuclear Displacements. 215
   3. Normal Modes and Normal Coordinates. 216
II. Phonon Correlation Function and Experimental Observables. 220
   4. General Discussion. 220
   5. Definition of Phonon Correlation Function. 221
   6. Physical Properties of the Phonon Correlation Function. 222
   7. Spectral Representation of the Phonon Correlation Function. 224
   8. Relation of the Phonon Correlation Function to Experiments. 227
   9. Ultrasonic Attenuation. 231
   10. Inelastic Neutron Scattering. 233
   11. Optical Measurements. 235
III. Phonon Green's Function. 240
   12. Definition of Phonon Green's Function. 240
   13. Functional Derivative Technique. 243
   14. Green's Function Equation. 245
   15. Formal Procedure of Calculating the Phonon Green's Function. 248
   16. Phonon Self-Energy Function. 253
   17. Dispersion and Damping of Excitations in Crystal. 257
   18. Perturbation Calculation. 260
IV. Explicit Calculation of Phonon Green's Function and Applications. 262
   19. Calculation of the Phonon Self-Energy Function. 262
   20. Ultrasonic Attenuation: General Formulation. 267
   21. Attenuation of Transverse Acoustic Phonons. 269
   22. Attenuation of Longitudinal Acoustic Phonons. 274
V. Phonon Boltzmann Equation. 279
   23. Lattice Transport Properties. 279
   24. Derivation of the Phonon Boltzmann Equation. 281
VI. Coupled Phonon-Photon System. 290
   25. Interaction of Lattice Vibrations with the Macroscopic Electromagnetic
      Fields. 290
   26. Dispersion of the Optical Phonons. 293
   27. Polariton Green's Function. 294
   28. Dielectric Susceptibility. 301

LAG–LAU

LAGALLY, M.G.: *see* WEBB, M.B.

LANG, Norton D. <inline>VOLUME 28, 1973</inline>

The Density-Functional Formalism and the Electronic Structure of Metal Surfaces

NORTON D. LANG

IBM Thomas J. Watson Research Center, Yorktown Heights, New York

I. Introduction 225
II. Density-Functional Formalism 228
   1. General Formulation 228
   2. Two Regimes of Approximation...................................... 232
   3. Thomas–Fermi Method and Its Extensions 235
   4. Many-Body Problem in One-Body Form 236
   5. Ground-State Surface Properties in Terms of Electron Density.......... 240
III. Electronic Structure of Unperturbed Metal Surfaces 242
   6. Replacing an Ionic Lattice by a Uniform Charge Background 243
   7. Planar Uniform-Background Model: General Discussion 246
   8. Planar Uniform-Background Model: Solutions........................ 251
   9. Planar Uniform-Background Model: Perturbative Introduction of Lattice
      Effects... 268
   10. Corrugated Uniform-Background Model 275
IV. Response of Metal Surfaces to Static Perturbing Charges 279
   11. Linear Response 281
   12. Nonlinear Response 289
   Note Added in Proof... 295
   Appendix: Surface-Energy Term σ_s 296

LAUDISE, R.A. <inline>VOLUME 12, 1961</inline>

Hydrothermal Crystal Growth

R.A. LAUDISE AND J.W. NIELSEN

Bell Telephone Laboratories, Murray Hill, New Jersey

I. Introduction.. 149

LAUDISE, R.A.

II. Apparatus.. 154
    1. Autoclaves: General.. 154
    2. Stress Analysis.. 156
    3. Materials Selection.. 158
    4. Closures.. 162
    5. Cans and Capsules... 174
    6. Temperature Measurement and Control................................ 178
    7. Pressure Measurement.. 179
    8. Safety.. 182
III. Quartz Crystallization.. 184
    9. General Considerations.. 184
   10. Thermodynamic Data... 184
   11. The Kinetics of Quartz-Crystal Growth............................ 184
   12. The Incorporation of Impurities During Quartz Crystal Growth..... 199
   13. The Nature of Hydrothermal Solutions Containing Silica........... 203
IV. Other Crystals Grown Hydrothermally..................................... 204
   14. Oxides.. 204
   15. Sulfides.. 213
   16. Halides... 215
   17. Miscellaneous... 216
V. Conclusions... 220
    Acknowledgments.. 222

LAX, Benjamin VOLUME 11, 1960

Cyclotron Resonance

BENJAMIN LAX AND JOHN G. MAVROIDES

Lincoln Laboratory, Massachusetts Institute of Technology, Lexington, Massachusetts

  I. Introduction.. 261
 II. Cyclotron Resonance of Free Charged Particles....................... 264
    1. Electrons... 264
    2. Ions... 273
III. Cyclotron Resonance of Carriers in Solids............................. 276
    3. Microwave Cyclotron Resonance in Semiconductors................... 280
    4. Phenomena in Metals.. 312
    5. Magnetoplasma Phenomena.. 343
    6. Infrared Cyclotron Resonance..................................... 355
 IV. New Developments.. 370
    7. Millimeter Cyclotron Resonance................................... 370
    8. Cross-Modulation... 374
    9. Cyclotron Resonance in Strained Germanium and Silicon............ 377
   10. Magnetoacoustic Resonance....................................... 379
   11. Cyclotron Resonance Generators and Amplifiers................... 383
  V. Summary and Future Prospects.. 388
    Acknowledgments... 399

LAZ–LEI

LAZARUS, David   VOLUME 10, 1960

Diffusion in Metals

DAVID LAZARUS

Department of Physics, University of Illinois, Urbana, Illinois

I. Introduction.. 71
II. Theory... 73
  1. Formal Diffusion Equations...................................... 73
  2. Temperature Dependence of Diffusion............................ 80
  3. Mechanisms for Diffusion....................................... 89
  4. Impurity Diffusion in Pure Metals.............................. 95
  5. Diffusion in Binary Alloys.................................... 101
III. Experimental Techniques.. 109
  6. Introduction.. 109
  7. Direct Measurements... 110
  8. Indirect Methods.. 112
IV. Experimental Results.. 114
  9. Introduction.. 114
  10. Semiempirical Models... 114
  11. Self-Diffusion in Pure Metals and Alloys..................... 115
  12. Chemical Diffusion... 119
  13. Grain-Boundary and Surface Diffusion......................... 121
  14. Diffusion in Stressed Metals................................. 123

LEIBFRIED G.   VOLUME 12, 1961

Theory of Anharmonic Effects in Crystals

G. LEIBFRIED AND W. LUDWIG

Institut für Reaktorwerkstoffe, Technische Hochschule, Aachen
and
Institut für Reaktorwerkstoffe der Kernforschungsanlage, Jülich

I. Introduction.. 276
II. Foundation of Lattice Dynamics.................................... 280
  1. Expansion of the Potential Energy.............................. 280
  2. Properties of the Coupling Parameters.......................... 283
  3. Normal Modes.. 302
III. Thermodynamics... 307
  4. The Free Energy... 307
  5. Calculation of Thermodynamic Functions........................ 315
  6. Thermodynamics in the Harmonic Approximation.................. 324

LEIBFRIED G.

IV. The Linear Chain.. 330
    7. The Free Energy.. 330
    8. The Thermal Equation of State................................ 336
    9. The Caloric Equation of State................................ 345
V. The Three-Dimensional Lattice.................................... 350
    10. The Free Energy... 350
    11. The Thermal Equation of State............................... 354
    12. The Caloric Equation of State............................... 360
VI. The Thermal Equation of State................................... 364
    13. General Parameters.. 364
    14. Central Forces 384
    15. van der Waals Crystals...................................... 389
    16. The Alkali Halides.. 395
VII. The Caloric Equation of State.................................. 410
    17. General Theory for High Temperatures........................ 410
    18. Discussion of Special Examples.............................. 413
    19. Debye Temperatures (Low Temperatures) 421
VIII. The Sound Velocity... 426
    20. Preliminary Remarks... 426
    21. The Linear Chain.. 429
    22. Three-Dimensional Lattices.................................. 433
IX. Appendix... 434
    23. Calculation of F_3.. 434
    24. Remarks on the Substitution of ω^2 by $\overline{\omega^2}$ 439
    25. Calculation of P and Q 441
    26. Calculation of Derivatives of the Coupling Parameters........ 442

LEVINE, J.D.: *see* DAVISON, S.G.

LEVY, Peter M. VOLUME 47, 1994

Giant Magnetoresistance in Magnetic Layered and Granular Materials

PETER M. LEVY

Physics Department New York University, New York, New York

I.   Introduction 367
II. Survey of Experimental Data 375

LEV–LEW

LEVY, Peter M.

III.	Theoretical Formulations of Transport	391
	1. Semiclassical Approach	392
	2. Quantum Theory	394
IV.	Sources of Magnetoresistance	397
	3. Origin of Giant Magnetoresistance	398
	4. Calculations of Scattering	401
V.	Quantum Theory of Magnetoresistance	404
	5. Hamiltonian	404
	6. Linear Response	409
	7. Simplifications due to Symmetry	416
	8. Limiting Cases	421
	9. Real-Space Formulation	426
	10. Momentum Space Formulation	431
	11. Effect of Superlattice Potential on Magnetotransport	440
	12. Global Conductivity for Inhomogeneous Structures	444
VI.	Discussion of Results	449
	13. General Trends	450
VII.	Future Directions	458
	14. Experimental Work	458
	15. Theoretical Advances	461
	Acknowledgments	462

LEWIS, H.W.

VOLUME 7, 1958

Wave Packets and Transport of Electrons in Metals

H.W. LEWIS

University of Wisconsin, Madison, Wisconsin

I. Introduction	353
II. Uncertainty Principle and Degeneracy	358
III. Degenerate Systems	362
IV. Electron-Impurity Scattering; Angular Momentum	365
1. One Dimension	365
2. Two Dimensions	370
3. Three Dimensions	373
V. Phonon Interactions	374
VI. Transport Phenomena	375

LIEBER, Charles M.   VOLUME 48, 1994

Preparation of Fullerenes and Fullerene-Based Materials

CHARLES M. LIEBER

CHIA-CHUN CHEN

Division of Applied Sciences and Department of Chemistry
Harvard University
Cambridge, Massachusetts

I.	Introduction	109
II.	Fullerene Clusters	112
	1. Formation of Fullerenes	112
	2. Separation of Pure Clusters	117
	3. Fullerene Characterization	121
	4. Isotopically Substituted Fullerenes	125
III.	Metal-Doped C_{60} Materials	128
	5. C_{60} Solids and Thin Films	128
	6. Considerations for Metal Doping	133
	7. Alkali Metal–Doped Fullerenes	134
	8. Alkaline Earth–Doped C_{60}	140
	9. New Directions	141
IV.	New Fullerene Building Blocks	142
	10. Endohedral Metal Fullerenes	142
	11. Carbon Nanotubes	145
V.	Concluding Remarks	147

LIEBER, Charles M.   VOLUME 48, 1994

Physical Properties of Metal-Doped Fullerene Superconductors

CHARLES M. LIEBER

ZHE ZHANG

Division of Applied Sciences and Department of Chemistry
Harvard University
Cambridge, Massachusetts

I.	Introduction	349
II.	Normal and Superconducting State Phenomenology	352
	1. Crystal and Electronic Structures	352
	2. Conductivity and Dimensionality	357
	3. Critical Fields, Penetration Depth, and Coherence Length	360
III.	Possible Mechanisms and Other Physical Properties	363
	4. Models for Fullerene Superconductivity	363
	5. Transition Temperature Versus Lattice Constant	368
	6. Energy Gap	371
	7. Phonons	378
	8. Isotope Effect	380
IV.	Concluding Remarks	384

LIEBERT, L.: *see* KELLER, P.

LINDE, Ronald K.: *see* DORAN, Donald G.

LIU, S.H. VOLUME 39, 1986

Fractals and Their Applications in Condensed Matter Physics

S.H. LIU

Solid State Division
Oak Ridge National Laboratory
Oak Ridge, Tennessee

I.	Introduction	207
	1. Mathematical Background	208
	2. Fractals in Condensed Matter Physics	212
II.	Measurements of Fractal Dimension	219
	3. Real-Space Measurements	219
	4. Small-Angle Light or Neutron Scattering	222
III.	Growth of Fractal Aggregates	227
	5. Computer Simulation of Growth	228
	6. Kinetic Theory of Aggregation	235
	7. Kinetic Theory of Dendritic Growth	239
	8. Experimental Studies	241
IV.	Diffusion on Fractal Networks	244
	9. Anomalous Diffusion	245
	10. Diffusion on the Sierpinski Gasket	249
	11. The ac Response of Rough Interfaces	253
V.	Elastic Properties of Fractal Networks	257
	12. Critical Elastic Threshold of Percolation Networks	257
	13. Spectral Dimension of Elastic Vibrations	263
VI.	Magnetic Ordering on Fractal Networks	267
	14. Magnetic Ordering of Ising Systems	267

LOBB, C.J.:   *see* TINKHAM, M.

LOW, William   SUPPLEMENT 2, 1960

Paramagnetic Resonance in Solids

WILLIAM LOW

Department of Physics, The Hebrew University, Jerusalem, Israel

Preface v

I. Introduction 1

    1. The Transition Groups 1
    2. Resonance Conditions 4
    3. Fine Structure 5

II. Theory of the Crystalline Field 8

    4. General Hamiltonian of the "Free Ion" 8
    5. Effect of Static Crystalline Field. Outline of Calculations 11
    6. Kramer's Theorem 34
    7. Jahn-Teller Effect 37
    8. The Spin Hamiltonian 39
    9. Angular Variation of the Spectrum 52
    10. Hyperfine Structure 58
    11. Direct Observation of Nuclear Magnetic Resonance via the Electron
        Spin Resonance 67
    12. The Determination of the Sign of D, A, Q', and γ 71

III. Paramagnetic Resonance Spectra of the Transition Elements in Single
    Crystals 76

    13. Paramagnetic Resonance Spectra of the Iron Group . . . 76
    14. Covalent Bonding in the Iron Group 97
    15. Optical Spectra of the Iron Group 103
    16. Energy Levels of Strongly Covalent Complexes 106
    17. The Cyanide Group 108
    18. The Palladium and Platinum Groups 111
    19. S-State Ions 113
        a. Introduction 113
        b. Energy Levels 114
        c. Spin Hamiltonian 116
        d. Experimental Results and Theoretical Considerations 118
        e. Anomalous Hyperfine Structure 122
    20. The Rare Earth Group 123
        a. Introduction 123
        b. Theoretical Aspects of the Paramagnetic Resonance Spectrum . . . 124
        c. Results of Paramagnetic Resonance Measurements 128
        d. Ions with an Even Number of Electrons 135

LOW

LOW, William

21. Paramagnetic Resonance Spectra of the Actinide Group 137
    a. Introduction. 137
    b. Uranyl Ions. 139
    c. Neptunyl Ions. 141
    d. Plutonyl Ions 144
    e. f^n Configurations. 145

IV. Relaxation Times and Line Width 149

22. Spin-Lattice Relaxation Time 149
23. Dipolar Broadening. 153
24. Exchange Interaction. 155
25. Methods of Measuring Relaxation Times 159
    a. Other Sources of Broadening 159
    b. Measurements of T_2 160
    c. Measurement of T_1 160
26. Solid State Maser. 162
    a. Pulsed Systems 165
    b. Adiabatic Fast Passage. 165
    c. Multilevel Scheme. 166

V. Paramagnetic Resonance of Color Centers and Defects in Crystals . . 170

27. Color Centers 170
    a. F Centers. 170
    b. V Centers. 173
28. Electron Spin Resonance of Irradiated Crystals (Other Than Alkali
    Halides). 175
    a. Spectrum of Hydrogen Atoms. 175
    b. Irradiated Nitrate Salts. 175
    c. Irradiated Diamonds, Quartz, and Other Crystals 176
    d. Organic Substances. 176
    e. Plastics. 177
    f. Resonance from Donors in Semiconductors 177

VI. Paramagnetic Resonance Spectrometers 180

29. Sensitivity Considerations. 180
    a. Frequency Considerations. 181
    b. Choice of Cavity. 182
    c. Modulation Frequency 184
    d. Detection Methods. 186
30. Practical Spectrometers. 190
    a. Simple Reflection or Transmission Spectrometer. 190
    b. Superheterodyne Spectrometer. 191
    c. Electron Nuclear Double Resonance (ENDOR) Spectrometer . . . 193

AUTHOR INDEX 195

SUBJECT INDEX 200

INDEX OF COMPOUNDS 211

LOW, W. VOLUME 17, 1965

Electron Spin Resonance of Magnetic Ions in Complex Oxides
Review of ESR Results in Rutile, Perovskites, Spinels, and Garnet Structures

W. Low

Department of Physics and National Magnet Laboratory, Massachusetts Institute of Technology, Cambridge, Massachusetts

AND

E.L. Offenbacher

Department of Physics, Temple University, Philadelphia, Pennsylvania

I. Introduction.. 136
II. Outline of ESR Specta in Inorganic Crystals......................... 138
   1. Energy Levels in Octahedral and Tetrahedral Symmetries............. 138
   2. Remarks on the Interpretation of Paramagnetic Resonance Results.... 141
   3. Remarks on the ESR of Rare Earth Spectra...................... 145
III. Single-Ion Contribution to Anisotropy Energy........................ 147
IV. The Spectra of Transition Elements in Simple Oxides (MgO, CaO, SrO, ZnO, and Al_2O_3).. 153
V. Rutile: TiO_2.. 154
   4. Crystal Structure... 154
   5. Interstitial Positions....................................... 156
   6. Individual Spectra.. 158
   7. General Conclusions... 161
VI. Perovskites.. 162
   8. Barium Titanate ($BaTiO_3$).................................. 164
   9. Strontium Titanate ($SrTiO_3$).............................. 167
   10. Miscellaneous Perovskite Structures........................... 176
VII. Spinels... 180
   11. Crystal Structure.. 180
   12. ESR Spectra.. 182
VIII. Garnets.. 185
   13. Crystal Structure.. 186
   14. ESR Spectra.. 188
   Appendix... 193

LUD–Mac

LUDWIG, G.W.   VOLUME 13, 1962

Electron Spin Resonance in Semiconductors

G.W. LUDWIG AND H.H. WOODBURY

General Electric Research Laboratory, Schenectady, New York

I. Introduction... 223
II. The Resonance Technique... 226
  1. The Spin Hamiltonian.. 226
  2. The Spin Resonance Spectrum................................... 231
  3. Experimental Techniques... 237
III. Resonance Studies in Silicon....................................... 243
  4. Shallow Donor Impurities.. 244
  5. Shallow Acceptor Impurities..................................... 259
  6. Transition Metal Ions... 263
  7. Impurity Pairs.. 273
  8. Radiation Damage Centers.. 280
IV. Resonance Studies in Other Semiconductors.......................... 286
  9. Germanium... 286
  10. Graphite and Diamond... 290
  11. Silicon Carbide.. 293
  12. Indium Antimonide and Gallium Phosphide........................ 294
  13. Zinc Sulfide and Related Semiconductors........................ 295
  Acknowledgments... 304

LUDWIG, W.: *see* LEIBFRIED, G.

LUNDQVIST, Stig: *see* HEDIN, Lars

LYNCH, R.W.: *see* DRICKAMER, H.G.

MacKINNON, A.: *see* MILLER, A.

MacLAUGHLIN, Douglas E. VOLUME 31, 1976

Magnetic Resonance in the Superconducting State

DOUGLAS E. MACLAUGHLIN

Department of Physics, University of California, Riverside, California

I. Introduction... 2
    1. Magnetic Resonance as a Probe of the Superconducting State......... 3
    2. Static Measurements: Frequency Shifts and Inhomogeneous Broadening. 4
    3. Dynamic Measurements: Spin-Lattice Relaxation.................. 5
II. Knight Shift in the Superconducting State........................... 5
    4. Early Experiments.. 5
    5. Yosida's Theory of the Spin Susceptibility...................... 7
    6. Spin-Orbit Scattering: The Ferrell–Anderson Mechanism............. 8
    7. Experiments on Aluminum, Tin, and Lead....................... 9
    8. Non-s Contributions: Vanadium, Niobium, and the V_3X Compounds .. 12
III. Nuclear Spin-Lattice Relaxation in BCS Superconductors............... 15
    9. The Theory of Hebel and Slichter............................. 15
    10. Energy-Gap Anisotropy: Measurements in Aluminum and Al-Based
        Alloys.. 18
    11. Phonon Damping and Energy-Gap Inhomogeneity: Measurements in
        Indium and In-Based Alloys.................................. 26
    12. Magnetic and Quadrupolar Relaxation Mechanisms................. 31
    13. Measurements in Other Superconductors........................ 32
IV. Quadrupole Coupling Constants in Superconductors.................... 33
V. Nuclear Magnetic Resonance in Type-II Superconductors............... 34
    14. Magnetic Field Distribution in the Mixed State.................. 35
    15. Spin-Lattice Relaxation Near H_{c2}: The Field-Induced Gapless State... 40
    16. Relaxation Far from H_{c2}: Diffusion to the Vortex Cores?............. 46
    17. Vortex Motion in the Resistive State........................... 49
VI. Transition-Metal Impurities in Superconductors...................... 50
    18. Paramagnetic Impurities: Pair Breaking and Spin-Lattice Relaxation.. 51
    19. Electron Spin Resonance of Paramagnetic Impurities.............. 54
    20. Resonant Nonmagnetic Impurity States......................... 57
VII. Thermodynamic Fluctuations in the Superconducting State............. 60
    21. Fluctuations of the Vortex Lattice............................. 60
    22. Fluctuations of the Order Parameter in Small Specimens........... 62
    23. Fluctuations Above T_0 in Bulk Superconductors.................. 68

Mcc

McCALDIN, James O.: *see* YU, Edward T.

McCLURE, Donald S. VOLUME 8, 1959

Electronic Spectra of Molecules and Ions in Crystals Part I. Molecular Crystals

DONALD S. MCCLURE

RCA Laboratories, Princeton, New Jersey

I. Introduction 1
II. Electronic States of Molecular Crystals 2
  1. Theory 4
    a. The Perturbation Problem 4
    b. The Use of Symmetry 5
    c. The Energy Levels 12
    d. The Interaction Potential 13
    e. Transition Probabilities 16
    f. Vibrational-Electronic States 19
    g. Effects of Anisotropy on the Spectra 23
    h. Separated Excitons 24
  2. Experimental Studies of the Spectra of Molecular Crystals 24
    a. Aromatic Hydrocarbons 26
    b. Other Molecules 38
    c. Mixed Crystals 39
    d. Triplet States 40
  3. Fluorescence 40
    a. Fluorescence in the Absence of Energy Transfer 41
    b. Fluorescence of Pure Crystals 41
    c. Energy Migration in Molecular Crystals 45
  4. Summary 46

McCLURE, Donald S.   VOLUME 9, 1959

Electronic Spectra of Molecules and Ions in Crystals
Part II. Spectra of Ions in Crystals

DONALD S. MCCLURE

RCA Laboratories, Princeton, New Jersey

I. Theory of Spectra of Ions in Crystals................................... 400
  1. Theory of Crystal Field Spectra..................................... 400
    a. A Simple Example.. 402
    b. Electrostatic Charge Distributions.............................. 406
    c. Energy Level Calculations....................................... 407
    d. Weak, Strong, and Intermediate Crystal Fields................... 408
    e. Spin-Orbit Coupling... 424
    f. Selection Rules... 430
    g. Intensities.. 433
  2. Molecular Orbital Theory and Charge Transfer Spectra............... 438
    a. The Need for Extending Crystal Field Theory..................... 438
    b. Formal Treatment of Molecular Orbital Theory................... 439
    c. Charge Transfer Processes...................................... 442
    d. Intensity of Charge Transfer Processes......................... 448
    e. The Change from Localized to Nonlocalized Orbitals.............. 449
    f. Conclusions.. 451
II. Experimental Studies of the Spectra of Ions in Crystals............... 452
  3. Spectra of the Lanthanides.. 453
    a. The Chemical and Physical Properties of the Lanthanides........ 453
    b. Spectra of the Free Ions....................................... 454
    c. Spectra of the Individual Ions in Crystals..................... 459
    d. Interaction of the Ion with the Lattice........................ 474
    e. Conclusions.. 476
  4. Spectra of Transition Metal Ions.................................. 476
    a. Chemical and Physical Properties of the Ions................... 476
    b. Spectra of Individual Ions..................................... 479
  5. Charge Transfer Spectra... 502
    a. Octahedral Coordination.. 504
    b. Tetrahedral Coordination....................................... 507
    c. Conclusions.. 511
  6. Spectra of the Post-Transition Metal Ions......................... 511
    a. The Transitions $s^2 \to sp$................................... 512
    b. The Transitions $d^{10} \to d^9 s$............................. 521
    c. The Transitions $d^{10}s \to d^{10}p$.......................... 523
    d. Spectra of Ion Pairs... 523
    e. Conclusions.. 524

McG

McGill, Thomas C.: *see* YU, Edward T.

McGreevy, Robert L. VOLUME 40, 1987

Experimental Studies of the Structure and Dynamics of Molten Alkali and Alkaline Earth Halides

ROBERT L. MCGREEVY

Clarendon Laboratory, University of Oxford,
Oxford, England

I. Introduction		247
II. Structure		249
	1. Experimental Techniques	249
	2. Experimental Results	253
	3. Computer Simulation and Theory	264
III. Dynamics: Inelastic Neutron Scattering		269
	4. Experimental Techniques	269
	5. Experimental Results	275
	6. Computer Simulation and Theory	297
	7. Discussion	299
IV. Dynamics: Light Scattering		301
	8. Experimental Techniques	301
	9. Experimental Results	303
V. Conclusions		322

McMURRY, S.:   *see* WEAIRE, D.

McQUEEN, R.G.:   *see* RICE, M.H.

MAHAN, G.D.   VOLUME 29, 1974

Many-Body Effects on X-Ray Spectra of Metals

G.D. MAHAN

Physics Department, Indiana University, Bloomington, Indiana

I. Introduction 75
   1. Introduction 75
II. Experiments and Early Theories 82
   2. Emission and Absorption 82
   3. Lithium and Beryllium 85
   4. Sodium, Magnesium, and Aluminum 88
   5. Liquids and Alloys 92
III. Independent Boson Model 94
   6. General Features 94
   7. Phonon Sidebands 98
   8. Collision Broadening of Atomic Spectral Lines 101
   9. Collective Excitations in Metals 103
  10. Shapes of Spectral Lines 107
  11. Infrared Divergence 111
  12. X-Ray Edge in Metals 115
  13. Nozieres and DeDominicis Solution 120
IV. Theory of the X-Ray Singularity 122
  14. Form of the Potential 122
  15. Hole Green's Function 124
  16. X-Ray Absorption 129
  17. Numerical Results 135
V. Epilogue 137

Good Thermoelectrics

G.D. MAHAN

Department of Physics and Astronomy, The University of Tennessee, Knoxville, Tennessee, and Solid State Division, Oak Ridge National Laboratory, Oak Ridge, Tennessee

I.	Introduction	82
II.	Devices	85
	1. Macroscopic Equations	85
	2. Device Equations	88
	3. Refrigerators	89
	4. Generators	90
	5. Cooling a Hot Source	91
III.	Bulk Semiconductors	91
	6. Power Factor	91
	7. Transport Theory	93
	8. Bipolar Contribution	98
	9. The B Factor	99
	10. Quality Factor	105
	11. Thermal Conductivity	106
	12. Electronegativity	109
IV.	Review of Materials	111
	13. Good Thermoelectrics	111
	14. Bi_2Te_3	112
	15. IV–VI Semiconductors	113
	16. TAGS	114
	17. Si–Ge	115
	18. Bi–Sb	116
	19. Skutterudites	117
V.	Narrow-Gap Semiconductors	119
	20. Low-Temperature Thermoelectrics	119
	21. $PtSb_2$	120
	22. $Hg_{1-x}Cd_xTe$	121
	23. Epilogue	123
	24. Quantum Wells	125
VI.	Metals	130
	25. Mixed-Valent Compounds	130
	26. Transition-Metal Compounds	135
	27. Rare-Earth Compounds	136
	28. $YbAl_3$	138
	29. Cerium Compounds	140
	30. Anderson Model	142
	31. Formula for the Seebeck Coefficient	148
VII.	Summary	152
	32. Upper Bounds on Z	152
	33. Conclusions	155

MANNEVILLE, P.: *see* DUBOIS-VIOLETTE, E.

MARADUDIN, A.A. SUPPLEMENT 3, 1963

Theory of Lattice Dynamics in the Harmonic Approximation

A.A. MARADUDIN

Westinghouse Electric Corporation, Pittsburgh, Pennsylvania

E.W. MONTROLL

International Business Machines Corporation, Yorktown Heights, New York

G.H. WEISS

*Institute for Fluid Dynamics and Applied Mathematics, University of Maryland,
College Park, Maryland*

Preface v

I. Introduction 1

II. Elements of the Theory of Lattice Dynamics 5

    1. Equations of Motion of a Vibrating Lattice 5
    2. The Cyclic Boundary Condition and Ledermann's Theorem . 21
    3. Normal Coordinates of a Lattice 29
    4. Historical Survey 38

III. Theory of the Vibrational Frequency Spectra of Solids 45

    1. Introduction 45
    2. Expressions for Frequency Spectra 53
    3. Singularities in Frequency Spectra 65
    4. Topological Basis for Singularities in Frequency Spectra . . 76
    5. Approximate Spectrum Calculations 84
    6. Vibrational Frequency Spectra of Lattices with Long-Range
       Interactions Between Ions 103

MAR

MARADUDIN, A.A.

IV. Calculations of Thermodynamic Functions without the Use of
   the Frequency Spectrum. 117
   1. Thirring's Expansion and Its Analytic Continuation . . . 117
   2. Houston's Method 121

V. The Effect of Defects and Disorder on Lattice Vibrations 129
   1. Introduction 129
   2. Rayleigh's Theorems 132
   3. Mathematical Preliminaries 136
   4. Lattice Models to be Studied 150
   5. Effects of Defects on the Vibrations of Monatomic Lattices . 159
   6. Effects of Defects on the Vibrations of Diatomic Lattices . . 179
   7. Disorder in Two-Component Lattices 188

VI. The Effects of Surfaces on the Vibrations of Crystal Lattices. . 213
   1. Size Effect in the Low Temperature Specific Heat of Small
      Particles 213
   2. Surface Vibration Modes in a Semi-Infinite Continuum . . 218
   3. Surface Vibration Modes in Discrete Lattices 219
   4. Interactions of Defects with Boundary Surfaces 227

VII. Scattering of X-Rays and Cold Neutrons by Lattice Vibrations. . 231
   1. Introduction 231
   2. Localization of an Atom about Its Equilibrium Position . . . 236
   3. Theory of the Scattering of X-Rays by Thermal Lattice
      Vibrations 242
   4. Experimental Determination of Frequency Spectra by X-Ray
      Methods 252
   5. The Determination of Lattice Vibrational Spectra by Slow
      Neutron Scattering Experiments 258
   6. Experimental Determination of Frequency Spectra by Neutron
      Diffraction Methods 264
   7. Time-Relaxed Position and Momentum Correlation Functions 273
   8. Time-Relaxed Momentum Correlation Functions and Irrever-
      sibility 281

REFERENCES 289

AUTHOR INDEX 299

SUBJECT INDEX 305

MARADUDIN, A.A.   SUPPLEMENT 3, (Second Edition) 1971

Theory of Lattice Dynamics in the Harmonic Approximation
Second Edition

A.A. MARADUDIN

Department of Physics, University of California, Irvine, California

E.W. MONTROLL

Department of Physics, University of Rochester, Rochester, New York

G.H. WEISS

National Institutes of Health, Bethesda, Maryland

I.P. IPATOVA

A.F. Ioffe Physico-Technical Institute, Leningrad, USSR

Foreword ix
Preface to the Second Edition xi
Preface to the First Edition xiii

I. Introduction 1
   References 4

II. Elements of the Theory of Lattice Dynamics 6
   1. Equations of Motion of a Vibrating Lattice 6
   2. The Cyclic Boundary Condition and Ledermann's Theorem . 38
   3. Normal Coordinates of a Lattice 46
   4. Correlation Functions and Green's Functions 57
   5. Historical Survey 68
   References 77

III. Applications of Group Theory to Lattice Dynamics 82
   1. Introduction 82
   2. The Symmetry of the Dynamical Matrix and Multiplier
      Representations 82
   3. Transformation Properties of the Eigenvectors of the
      Dynamical Matrix 94
   4. Additional Degeneracies Due to Time Reversal Symmetry . 106
   References 128

MAR

MARADUDIN, A.A.

IV. Theory of the Vibrational Frequency Spectra of Solids 129

    1. Introduction 129
    2. Expressions for Frequency Spectra 137
    3. Singularities in Frequency Spectra 150
    4. Topological Basis for Singularities in Frequency Spectra . . 158
    5. Approximate Spectrum Calculations 166
    References 186

V. Calculations of Thermodynamics Functions without the Use
of the Frequency Spectrum 189

    1. Thirring's Expansion and Its Analytic Continuation . . . 189
    2. Houston's Method 193
    References 198

VI. Dynamical Theory of Ionic Crystals 200

    1. Introduction 200
    2. The Macroscopic and Lorentz Fields in an Ionic Crystal;
       Ewald's Method 201
    3. The Rigid Ion Model 221
    4. Polarizable Ions 234
    5. Phenomenological Theory 244
    6. Polaritons 257
    7. Optical Activity 269
    8. Vibrational Frequency Spectra of Crystals with Long Range
       Interactions between Ions 288
    References 297

VII. Scattering of X Rays and Cold Neutrons by Lattice Vibrations . . . 300

    1. Introduction 300
    2. Localization of an Atom about Its Equilibrium Position . . 304
    3. Theory of the Scattering of X Rays by Thermal Lattice
       Vibrations 311
    4. Experimental Determination of Frequency Spectra by X-Ray
       Methods 320
    5. The Determination of Lattice Vibrational Spectra by Slow
       Neutron Scattering Experiments 325
    6. Experimental Determination of Frequency Spectra by
       Neutron Diffraction Methods 334
    7. Van Hove's Time-Relaxed Pair Distribution Function . . 343
    References 348

MARADUDIN, A.A.

VIII. The Effect of Defects and Disorder on Lattice Vibrations 353

   1. Introduction 353
   2. Rayleigh's Theorems 356
   3. The Equations of Motion of a Crystal Containing an Isolated
      Impurity Atom, and Their Solution 360
   4. The Frequency Spectrum of a Perturbed Crystal 384
   5. The Vibrational Green's Function for an Imperfect Crystal . 396
   6. Time-Dependent Displacement and Momentum Correlation
      Functions 399
   7. The Use of Symmetry and Group Theory in the Solution of
      Lattice Dynamical Defect Problems 402
   8. Examples 428
   9. Expansions in Powers of the Impurity Concentration . . . 476
  10. Disordered Crystals 492
      References 514

IX. The Effects of Surfaces on the Vibrations of Crystal Lattices . . . 520

   1. Introduction 520
   2. Surface Vibration Modes 521
   3. The Green's Function for a Crystal Slab 564
   4. The Surface Contribution to the Specific Heat of a Crystal . . 582
   5. Scattering from Crystal Surfaces 595
   6. The Interaction of Defects with Crystal Surfaces 620
   7. Scattering of Phonons by Crystal Surfaces 629
      References 634

X. Statistical Mechanical Properties of Systems of Coupled Harmonic
Oscillators 637

   1. Introduction 637
   2. Brownian Motion 637
   3. Poincaré Cycles 649
   4. Nonequilibrium Processes in Disordered Crystals 657
   5. Energy Flow in Crystals 664
      References 665

Appendix. The Role of the Conduction Electrons in Determining
      the Phonon Frequencies of Metals 667

      References 672

Author Index 673

Subject Index 683

MAR

MARADUDIN, A.A.   VOLUME 18, 1966

Theoretical and Experimental Aspects of the Effects of Point Defects and Disorder on the Vibrations of Crystals—1

A.A. MARADUDIN

Westinghouse Research Laboratories, Pittsburgh, Pennsylvania

" . . . *and now remains*
That we find out the cause of this effect,
Or rather say, the cause of this defect,
For this effect defective comes by cause;

—*Wm. Shakespeare, hamlet*
Act II, Scene2

I. Introduction.. 274
II. Dynamical Properties of Isolated Impurity Atoms...................... 279
III. Dynamical Properties of Disordered Crystals......................... 298
    1. The Frequency Spectrum of a Disordered Crystal.................... 298
    2. The Specific Heat of a Disordered Crystal........................ 320
IV. Electrical Properties of Crystals and Defect Lattice Dynamics........... 324
    3. Electrical Resistivity... 324
    4. On the Possibility of Observing Localized and Resonance Modes in Super-
       conducting Tunneling Characteristics............................. 326
V. Effects of Point Defects on the Lattice Thermal Conductivity of Insulators... 328
    5. Introduction.. 328
    6. The Effects of Isotopic Impurities on the Low-Temperature Thermal
       Conductivity of Insulators....................................... 329
    7. Evidence for Resonant Scattering of Phonons by Point Defects........ 333
    8. Theory... 335
VI. The Scattering of Neutrons by Lattice Vibrations in Imperfect Crystals...... 345
    9. Theory... 346
    10. Experimental Results... 361
VII. The Mössbauer Effect for an Impurity Nucleus........................ 367
    11. Theory.. 367
    12. Experimental Results... 385
VIII. Optical Analogs of the Mössbauer Effect............................ 392
    13. Theory of Optical Spectra Due to Impurity Centers in Crystals........ 392
    14. Experimental Work... 410

MARADUDIN, A.A.   VOLUME 19, 1966

Theoretical and Experimental Aspects of the Effects of Point
Defects and Disorder on the Vibrations of Crystals—2

A.A. MARADUDIN

Westinghouse Research Laboratories, Pittsburgh, Pennsylvania

I. Impurity-Induced Infrared Lattice Absorption in Crystals   2
   1. Theory   3
   2. Experimental Results for Ionic Crystals 12
   3. Impurity-Induced One-Phonon Absorption in Homopolar Crystals 28
   4. Impurity-Induced First-Order Raman Scattering of Light by Crystals 37
II. Anharmonic Effects in Impurity-Induced One-Phonon Infrared Lattice
   Absorption 51
   5. Introduction 51
   6. Lifetime of a Localized Vibration Mode 55
   7. Side Bands to the Localized-Mode Peak 63
   8. Overtones of the Localized-Mode Peak in the Absorption Spectra of Ionic
      Crystals Containing U-Centers 69
   9. Some Effects of Anharmonicity on the Mössbauer Effect for an Impurity
      Nucleus 75
III. Localized Modes and Spin-Lattice Interactions 83
   10. Theory of Spin-Lattice Relaxation through Localized Modes 85
   11. Experimental Results 95
IV. The Surface of a Crystal Considered as a Defect 100
   12. Introduction 100
   13. Studies of the Dynamical Properties of Surface Atoms by Low-Energy
      Electron Diffraction 100
   14. The Mössbauer Effect for a Surface Atom 103
   15. Theory 107
   16. Surface Vibration Modes 117
   17. The Surface Contribution to the Specific Heat of a Crystal 121
   18. Interaction of Defects and Crystal Surfaces 130
V. Conclusion 133

MAR

MARCH, N.H.: *see* CALLAWAY, J.

MARKHAM, Jordan, J.   SUPPLEMENT 8, 1966

F–Centers in Alkali Halides

JORDAN J. MARKHAM

*Department of Physics, Illinois Institute of Technology, Technology Center,
Chicago, Illinois*

PREFACE vii

Chapter I. The Optical Properties of the F-Center

1. Introduction 1
2. Electron Injection in Alkali Halides 3
3. The Optical Properties of the F-Center 7
4. Smakula's Equation 26
   Appendix 34

Chapter II. The Production of F-Centers

5. Production of F-Centers by Irradiation 37
6. Additive Coloration 58

Chapter III. The Photoconductivity of the F-Center

7. Introduction 65

Chapter IV. Other Optical Properties of F-Centers

8. The Luminescence 89
9. The Bleaching of the F-Center at Low Temperature . . . 96
10. The α- and β-Bands 106
11. The Thermal Ionization of an F-Center 113
12. An Energy Level Diagram for the F-Center 124
13. Further Considerations of Bleaching and Photoconductivity . . 127

Chapter V. Some Optical Effects Related to the F-Band

14. Introduction 135

MARKHAM, Jordan J.

Chapter VI. Electron Paramagnetic Resonance in a Medium without Nuclear Spins

15. Introduction 159
16. Summary of Quantum Mechanical Effects 161
17. Solution of Dynamic Equations for an Isolated System . . . 181
18. General Theory of Paramagnetic Resonance (Quantum Concepts) 192
19. Theory of the g-Factor as Applied to the F-Center 201
    Appendix 206

Chapter VII. Effect of the Nuclear Spins on the Electron Paramagnetic Resonance

20. Hyperfine or Inhomogeneous Broadening 221
21. Saturation and Hyperfine Broadening 233
22. The Electron Nuclear Double Resonance (ENDOR) Technique 243

Chapter VIII. Experimental Results on the Magnetic Properties of the F-Center

23. The Static Paramagnetic Susceptibility 255
24. Electron Paramagnetic Resonance 258

Chapter IX. The Theory of the F-Center Wave Functions

25. Introduction: The Model 289
26. The Aim of the Theory 293
27. Fundamental Assumptions Regarding the Calculation of Wave Functions at Point Imperfections 297
28. The F-Center Calculations 309

Chapter X. Strong Electron-Phonon Interactions

29. Crystal and Local Modes 331
30. Configurational Coordinate Diagrams 353
31. The Franck-Condon Principle and the Shape of an Absorption and an Emission Band 362
32. Nonradiative Transitions 377
33. Experimental Results 378
    Appendix 385

AUTHOR INDEX 389

SUBJECT INDEX 395

MAR–MEI

MARTIN, Georges   VOLUME 50, 1996

Driven Alloys

GEORGES MARTIN AND PASCAL BELLON

*Section de Recherches de Métallurgie Physique, DTA/CEREM/DECM,CEA Saclay,
Gif-sur-Yvette, France*

I. Introduction 189
II. Typical Examples of Phase Transformations in Driven Alloys. 191
    1. Alloys under Irradiation 191
    2. Alloys under Plastic Deformation 202
    3. Thin Films Grown from Vapor Phase 216
III. Phenomenological Theory 221
    4. Basic Ideas 221
    5. Heuristic Model 222
    6. Atomistic-Level–Based Models. 230
IV. Typical Predictions and Experimental Assessments 249
    7. A2-B2 Order-Disorder Transition under Forcing 250
    8. A1-L1$_2$ Order-Disorder Transition under Forcing 287
    9. Ni$_4$Mo Alloy under Irradiation 295
    10. Surface Segregation under Irradiation 300
    11. Alloys under Sustained Shearing 301
V. Conclusion 319
Acknowledgments. 319
Appendix A: Solid Solution Evolution in the Presence of Nonconservative
    Point Defects. 320
Appendix B: Lyapunov, Fokker-Planck, and Kubo Potentials 321
Appendix C: Nonequilibrium Grand Canonical Ensemble 327

Mavroides, John G.:   *see* LAX, Benjamin

MEIJER, Paul H.E.:   *see* HERZFELD, Charles M.

MENDELSSOHN, K.   VOLUME 12, 1961

The Thermal Conductivity of Metals at Low Temperatures

K. MENDELSSOHN AND H.M. ROSENBERG

The Clarendon Laboratory, Oxford, England

MICHEL, J.: *see* KIMERLING, L.C.

MIL

MILLER, A. VOLUME 36, 1981

Beyond the Binaries—The Chalcopyrite and Related Semiconducting Compounds

A. MILLER

Physics Department, North Texas State University, Denton, Texas

A. MacKINNON

Physikalisch-Technische Bundesanstalt, Braunschweig, Federal Republic of Germany

AND

D. WEAIRE

Physics Department, Heriot-Watt University, Edinburgh, Scotland

MITRA, Shashanka, S.   VOLUME 13, 1962

Vibration Spectra of Solids

SHASHANKA S. MITRA

Ontario Research Foundation, Toronto, Ontario, Canada

I. Introduction   2
   1. Separation of Internal Energy of a Molecule   2
   2. Vibrational Energy of a Molecule   2
   3. Raman Spectra   4
   4. Normal Vibrations   5
   5. Spectra in the Condensed Phases   6
   6. Experimental Techniques   9
II. Symmetry and Normal Vibrations in a Solid   11
   7. Fundamental Modes of Vibration of a Crystal   11
   8. Classification of Crystal Fundamentals   12
   9. Application of Group Theory   13
  10. Selection Rules   16
  11. Normal Modes of Oscillation in Brucite   17
  12. Polarization of Infrared Bands   21
  13. Depolarization and Intensity of Raman Lines   26
  14. Overtones and Combinations   29
III. Born-von Kármán Theory of Lattice Vibration   30
  15. Monatomic Chain   31
  16. The Linear Diatomic Chain   32
  17. Vibration of Cubic Lattices   35
  18. Raman's Theory and Finite Space Group Analysis   38
  19. Lattice Specific Heat   41
IV. Low-Frequency Librations in Molecular Crystals   45
  20. Rotatory Lattice Modes (Spectroscopic Evidence)   45
  21. Nature of Molecular Rotation in Solids   47
  22. Libration-Vibration Combination Spectrum   49
V. Infrared and Raman Spectra   53
  23. Alkali Halides   53
  24. Diamond Structure   59
  25. Lattice Infrared Spectra of Semiconductors   63
  26. Hydroxides of Alkaline Earths and Alkali Metals   66
  27. Ice   78
  Acknowledgments   80

MIT–MON

MĬTROVIĆ, Bozidar: *see* ALLEN, Philip B.

MIYAKE, Satoru J.: *see* KUBO, Ryogo

MOLLWO, E.: *see* HEILAND, G.

MONTGOMERY, D.J.   VOLUME 9, 1959

Static Electrification of Solids

D.J. MONTGOMERY

Michigan State University, East Lansing, Michigan

I. Foreword	139
II. Historical and Technical Background	141
III. Experimental Studies	144
1. Relevant Variables	144
2. Experimental Techniques	145
3. Results—Qualitative and Semiquantitative	160
4. Results—Quantitative	163
IV. Theoretical Interpretations	174
V. Status of the Problem	184
5. Comparison of Theory and Experiment	184
6. Scientific Consequences of the Theory	190
7. Technological Consequences of the Theory	191
8. Directions of Future Work	192
Appendix A. Chemical Nature of Certain Polymers	194
Appendix B. Redistribution of Charge on the Surface of a Dielectric	194

MONTROLL, E.W.:   *see* MARADUDIN, A.A.

MOSS, S.C.:   *see* AXE, J.D.

MUTO, Toshinosuke   VOLUME 1, 1955

The Theory of Order-Disorder Transitions in Alloys

TOSHINOSUKE MUTO

Institute of Science and Technology, University of Tokyo, Tokyo, Japan

AND

YUTAKA TAKAGI

Tokyo Institute of Technology, Tokyo, Japan

I. Introduction.. 194
   1. Direct Experimental Evidences for the Ordered Arrangement of Atoms
      in Alloys.. 194
   2. Order Parameters.. 197
   3. Measurements of the Order Parameters by the X-Ray Diffraction Method 200
   4. Main Problems Connected with the Order-Disorder Transitions........ 203
II. Theory of the Order-Disorder Transition in Alloys.................... 205
   5. General Survey.. 205
   6. Statistical Theory of Equilibrium Properties of the Order-Disorder
      Transition. I 207
      a. General Formulation with Introductory Remarks................. 207
      b. The Bragg-Williams Approximation.......................... 210
      c. Kirkwood's Method of Solution............................. 214
   7. Statistical Theory of Equilibrium Properties of the Order-Disorder
      Transition. II 221
      a. Bethe's Method of Solution................................. 221
      b. Quasi-chemical Method of Fowler-Guggenheim and Method of Takagi 227
      c. Exact Method of Solution Due to Kramers-Wannier and Onsager.... 237
      d. Comparison with the Experiments and Discussions................ 243
   8. Effect of Lattice Vibrations and Propagation of Order............... 246
   9. Kinetics of the Order-Disorder Transitions......................... 250

MUTO, Toshinosuke

III. Influence of Order on Electrical Properties............................ 252
  10. Theoretical Considerations... 252
  11. Change of Resistivity.. 253
  12. Change of Hall Coefficient... 261
IV. Influence of Order-Disorder on Magnetic Properties.................... 261
  13. Effect of Order on Diamagnetic and Paramagnetic Alloys............. 261
  14. Effect of Order on Ferromagnetic Properties........................ 262
    a. Saturation Moment of Ordered Alloys............................ 262
    b. Magnetostriction.. 266
    c. Magnetic Anisotropy... 266
    d. Coercive Force.. 267
    e. Curie Temperature... 268
  15. Order-Disorder in Ferromagnetic Alloys............................ 268
V. Influence of Order on Mechanical Properties........................... 272
  16. Change of Elastic Constants.. 272
  17. Change of Plastic Properties....................................... 276
  18. Change of Volume on Ordering....................................... 280
VI. Concluding Remarks.. 281

NAGAMIYA, Takeo   VOLUME 20, 1967

Helical Spin Ordering—1 Theory of Helical Spin Configurations

TAKEO NAGAMIYA

Department of Material Physics, Faculty of Engineering Science, Osaka University
Toyonaka, Japan

Introduction.. 306
I. Elementary Theory of Helical Spin Ordering.......................... 307
  1. Simple Helical Spin Ordering at Absolute Zero..................... 308
  2. Molecular Field Theory for Finite Temperature.................... 310
II. Spin Waves in the Screw Structure................................... 312
III. Effect of Anisotropy Energy on Spin Configuration.................... 316
  3. Uniaxial Anisotropy Energy with an Easy Axis; Oscillation in the Longitudinal Component... 317
  4. Uniaxial Anisotropy Energy with an Easy Axis; Oscillation in the Transverse Component.. 319
  5. Anisotropy Energy of Twofold, Fourfold, and Sixfold Symmetry with an Easy Plane... 323
  6. Anisotropy Energy Having an Easy Cone............................ 328

NAGAMIYA, Takeo

NEL–NEU

NELSON, David R.   VOLUME 42, 1989

Polytetrahedral Order
in Condensed Matter

DAVID R. NELSON

Lyman Laboratory of Physics
Harvard University
Cambridge, Massachusetts

FRANS SPAEPEN

Division of Applied Sciences
Harvard University
Cambridge, Massachusetts

I.	Introduction	1
II.	Liquids, Glasses, and Polytetrahedral Crystals	5
	1. Clusters	5
	2. Polytetrahedral Crystals	10
	3. Liquids	13
	4. Glasses	15
III.	Statistical Geometry	22
	5. Frustration in Two Dimensions	22
	6. Three Dimensions	29
IV.	Computer Simulations, the Glass Transition, and Frustration in Binary Mixtures	44
	7. Measurements of Short-Range Icosahedral Order in Liquids	44
	8. A Scenario for the Glass Transition	52
	9. Frustration and Glass Formation in Binary Alloys	55
V.	Statistical Mechanics of Polytetrahedral Order	57
	10. Statistical Estimate of the Crystal-Melt Interfacial Tension	57
	11. Order Parameter Theory of Polytetrahedral Statistical Mechanics	62
VI.	Quasicrystals	75
	12. Experiments and Theoretical Background	75
	13. Theory of Polytetrahedral Quasicrystals	82
	Appendix: The Statistical Honeycomb Model	88
	Acknowledgments	90

NEUMANN, D.A.:   *see* AXE, J.D.

NEWMAN, R. VOLUME 8, 1959

Photoconductivity in Germanium

R. NEWMAN AND W.W. TYLER

General Electric Research Laboratory, Schenectady, New York

I. Introduction	50
II. Electronic Structure of Germanium and Impurities	52
1. Band Structure of Germanium	52
a. Valence Band	52
b. Conduction Band	53
2. Nature of Impurity States	54
III. Photoconductivity Theory	56
3. Details of the Absorption Process	56
4. Nature of the Recombination Process	59
5. The Photoconductive Equations	62
IV. Preparation and Electrical Properties of Doped Germanium Crystals	64
6. Purity and Perfection of Undoped Crystals	65
7. Mobility	66
8. Summary of Deep Level Impurities in Germanium	70
9. Solubility and Diffusivity of Impurities	75
V. Experimental Studies of Photoconductivity	78
10. Extrinsic Photoconduction	79
a. Hydrogen-like Impurities	79
b. Impurities with Deep Levels	81
11. Steady State Intrinsic Photoconduction	92
12. Transient Measurements, Intrinsic Excitation	94
13. Complex Effects	101
a. Quenching	101
b. Superlinearity	103
VI. Germanium as an Infrared Detector	104
14. Intrinsic Photoconductivity ($\lambda < 1.8\,\mu$)	105
15. Impurity Photoconductivity	105
a. Absorption of Light	105
b. Variations in Spectral Sensitivity	106
c. Influence of Background Radiation	106
VII. Acknowledgments	107

NIC–NIE

NICHOLS, D.K. VOLUME 18, 1966

Energy Loss and Range of Energetic Neutral Atoms in Solids

D.K. NICHOLS AND V.A.J. VAN LINT

General Atomic Division of General Dynamics, John Jay Hopkins Laboratory for Pure and Applied Science, San Diego, California

I. Introduction	1
II. The Interatomic Potentials	3
1. One-Atom Potential Calculations	4
2. Two-Atom Potential Calculations	10
III. Scattering Calculations	14
IV. Range-Energy Calculations	20
3. Continuous Methods	21
4. Statistical Methods	23
V. End Effects	32
5. Displacement Collisions	32
6. Dynamic Model of the Lattice	34
7. Focusing, Sputtering, and Channeling	35
VI. Elastic versus Inelastic Collisions	36
VII. Range-Energy Experiments	38
8. Recoil Atom Measurements	40
9. Ion Beam Experiments	42
VIII. Scattering Experiments and Empirical Potentials	47
IX. Channeling and Superchanneling	51
Appendix	54

NIELSEN, J.W.: *see* LAUDISE, R.A.

NILSSON, P.O.   VOLUME 29, 1974

Optical Properties of Metals and Alloys

P.O. NILSSON

Department of Physics, Chalmers University of Technology, Gothenburg, Sweden

I. Introduction	139
1. Content of the Review	139
2. Definitions of Optical Constants	141
3. Theory for Optical Absorption	142
II. Experimental	145
4. Experimental Methods	145
5. Accuracy of Data	147
III. Interband Transitions	149
6. Pure Elements	149
7. Alloys	190
8. Modulation Experiments	211
IV. Intraband Transitions	218
V. Plasma Effects	223
9. Infinite Medium	223
10. Semiinfinite Medium	225
11. Thin Film	229

NOVEY, Theodore B.:   *see* HEER, Ernst

NUR

NURMIKKO, A.V.   VOLUME 49, 1995

Physics and Device Science in II–VI Semiconductor Visible Light Emitters

A.V. NURMIKKO

*Division of Engineering and Department of Physics, Brown University,
Providence, Rhode Island*

R.L. GUNSHOR

School of Electrical Engineering, Purdue University, West Lafayette, Indiana

I. Overview of Wide Band-Gap II–VI Semiconductors and Blue-Green
Light Emitters 205
II. Electronic States in Wide Band-Gap II–VI Quantum Wells 210
   1. Electronic Confinement and Band Offsets: General Comments 210
   2. ZnCdSe/ZnSe Quantum Wells: A Case Study of
   Two-Dimensional Excitons 217
   3. Tellurium Isoelectronic Centers in ZnSe and Exciton–Phonon Interaction . . 225
   4. Toward Quantum Wires and Dots 229
III. Doping and Transport 233
   5. Overview and Current Status 233
   6. *p*-Type Doping of ZnSe and Related Compounds 235
   7. Development of Low-Resistance Contacts to *p*-ZnSe 241
IV. Blue-Green LEDs and Diode Lasers 244
   8. Light-Emitting Diodes 245
   9. Diode Lasers 250
   10. Degradation and Defects 259
   11. Surface-Emitting Lasers: First Steps 264
V. Physics of Gain and Stimulated Emission in ZnSe-Based Quantum Well Lasers . 267
   12. Overview Comments 268
   13. Excitonic Processes in Optically Pumped Lasers at Low Temperature 269
   14. Gain Spectroscopy of Blue-Green Diode Lasers 277
VI. Summary 281

NUSSBAUM, Allen   VOLUME 18, 1966

Crystal Symmetry, Group Theory, and Band Structure Calculations

ALLEN NUSSBAUM

Electrical Engineering Department, University of Minnesota, Minneapolis, Minnesota

I. Introduction.. 165
II. The Algebra of Groups... 168
   1. The Fundamental Concepts of Group Theory..................... 168
   2. An Introduction to Group Representations..................... 171
   3. The Crystallographic Point Groups........................... 176
   4. The Character System of a Group............................. 182
III. The LCAO or Tight-Binding Approximation........................ 185
   5. A Review of the Fundamental Principles....................... 185
   6. The Method of Slater and Koster............................. 191
IV. The Band Structure of Beta-Brass............................... 201
   7. A General Description of Beta-Brass.......................... 201
   8. Character System for the Holohedral Cubic Group O_h........ 202
   9. The Free-Electron Approximation............................. 204
   10. The Tight-Binding Approximation for Beta-Brass............. 215
V. The Band Structure of Tellurium................................ 225
   11. The Space Group for Tellurium.............................. 225
   12. The Free-Electron Bands and the Compatibility Relations.... 229
   13. Space Groups.. 236
   14. The Nearly Free and Tight-Binding Approximations for Tellurium..... 245
VI. The Band Structure of Silicon and Germanium................... 249
   15. The OPW Method.. 249
   16. The Symmetry of the Diamond Lattice....................... 255
   17. Energy Bands for Silicon and Germanium.................... 261
   18. Spin and Double Groups.................................... 268

OFFENBACHER, E.L.: *see* LOW, W.

OKA–OSG

OKANO, K.: *see* SAITO, N.

OSGOOD, Richard M., Jr. VOLUME 51, 1997

Image States on Single-Crystal Metal Surface

RICHARD M. OSGOOD, JR., AND XIAOYI WANG

Columbia Radiation Laboratory, Columbia University,
New York, New York

I.	Introduction	2
	1. Image States and Condensed-Matter Physics	2
	2. New Measurement Techniques for Excited-State Photoemission	3
	3. Outline of This Chapter	5
II.	Our Theoretical Understanding of Image States	6
	4. The Simple Picture	6
	5. Two-Band Phase-Accumulation Theory of Image States	8
	6. Many-Body Theory for Image States	16
	7. Lifetime of Image States	19
	8. Image Resonances	21
III.	Experimental Probes of Image-State Properties	24
	9. Inverse Photoemission	25
	10. Scanning Tunneling Microscopy	25
	11. Two-Photon Photoemission	26
IV.	Experimental Measurements of the Stationary Properties of Image States on Bare Single-Crystal Surfaces	35
	12. Binding Energies	35
	13. Movement Parallel to the Image Plane	39
	14. Experimental Observations of Image Resonances	41
V.	Measurements of the Dynamic Properties of Image States	43
VI.	Image States on Ferromagnetic Surfaces	46
VII.	Image States and Surface Conditions	50
	15. Temperature-Induced Changes in Image States	51
	16. Adsorbate-Induced Changes in Image-State Properties	53
	17. Image States and Metallic Quantum Wells	57
	18. Image Electrons in the Presence of Dielectric Layers on Metal Surfaces	60
VIII.	Lateral Quantum Confinement of Surface Electrons	63
	19. Scattering from Individual Surface Impurity Atoms	64
	20. Confinement on Adsorbate Islands	68
	21. Lateral Superlattices on Stepped Surfaces	70
	22. Height-Dependent Electronic Behavior on Stepped Surfaces	75
IX.	Summary	78

Nuclear Magnetic Resonance

G.E. PAKE

Washington University, St. Louis, Missouri

Introduction.. 1
I. Magnetic Resonance.. 2
   1. The Nuclear Magnet... 2
   2. Nuclear Spin Equilibrium in a Constant External Field.............. 7
   3. Magnetic Resonance as Described by Rotating Coordinate Methods.... 8
   4. Interactions Involving the Spins; Nuclear Magnetic Relaxation........ 12
      a. Spin-Spin Interactions... 12
      b. Spin-Lattice Interactions...................................... 15
   5. The Bloch Formulation... 19
II. Methods for Observing Resonance and Measuring the Relaxation Times... 25
   6. Experimental Methods for Observation of Continuously Excited Nuclear
      Resonance... 25
      a. The Bridge Method... 25
      b. Signal-to-Noise Considerations; Narrow Banding................. 29
      c. Variable Frequency Spectrometers............................. 31
   7. Free Induction Decays and Spin Echoes............................ 32
   8. Measurement of T_1 by Methods Other than Echoes.................. 38
III. Resonance Widths, Splittings, and Shifts.............................. 39
   9. Dipolar Broadening in Rigid Lattices.............................. 39
   10. Influence of Nuclear Motion on Dipolar Broadening................. 44
   11. An Example: Proton Magnetic Resonance in 1,2-Dichloroethane....... 47
   12. Nuclear Electric Quadrupole Interaction........................... 50
   13. Resonance Shifts... 54
      a. The Knight Shift... 54
      b. The Chemical Shift.. 57
   14. Electron-Coupled Spin-Spin Interactions........................... 58
      a. The Discovery and Explanation of the Coupling................. 58
      b. Production of Very Homogeneous Magnetic Fields................ 63
      c. Proton Exchange and Forced-Flip Narrowing.................... 63
IV. Spin-Lattice Relaxation.. 67
   15. The BPP Theory of Relaxation in Liquids.......................... 67
   16. The Contribution of Lattice Vibrations to $1/T_1$ for Crystals.......... 73
   17. Nuclear Relaxation by Other Nuclear Motions in Solids.............. 77
   18. Nuclear Relaxation by Conduction Electrons....................... 82
   19. Electric Quadrupole Relaxation Mechanisms........................ 84
   20. Electron Spin Resonance.. 84
   21. The Overhauser Polarization of Nuclei............................. 87

PAL–PAR

PALM, J.:   *see* KIMERLING, L.C.

PARKER, R.L.   VOLUME 25, 1970

Crystal Growth Mechanisms: Energetics, Kinetics, and Transport

R.L. PARKER

National Bureau of Standards, Washington, D.C.

Introduction 152
I. Thermodynamics of Phase Changes: Phenomenology of the Problem 155
   1. Introduction 155
   2. Thermodynamics of Bulk Phase Equilibrium 156
   3. Thermodynamics of Interfaces 158
   4. Nonequilibrium Thermodynamics 160
II. Statistical Mechanics of Crystal Growth: Theoretical Physics of the Problem . 161
   5. Equilibrium Statistical Mechanics 161
   6. Applications of Equilibrium Statistical Mechanics to Phase Changes 162
   7. Nonequilibrium Statistical Mechanics 170
III. Stefan Problem: Mathematical Physics of Crystal Growth Transport Processes 170
   8. Introduction 170
   9. Stefan Problem without Complications: Various Shapes 172
   10. Stefan Problem with Complications: Various Shapes 185
   11. Some Approximate Solutions to the Stefan Problem 195
   12. Some Questions Regarding Stefan Problems 197
IV. Nucleation 197
   13. Introduction 197
   14. Classical Nucleation 198
   15. Some Modern Developments in Nucleation 204
V. Interface Structures and Interface Kinetics 208
   16. Interface Structures in Equilibrium 208
   17. Vapor Phase Growth Interface Kinetics: Theoretical and Experimental . . 226
   18. Solution Growth Interface Kinetics 239
   19. Melt Growth Interface Kinetics 245
   20. Kinematic Theory of Crystal Growth 250
VI. Morphological Stability 253
   21. Introduction 253
   22. Morphological Stability Problem without Complications: Various Shapes 255
   23. Stability Problem with Complications 262
   24. Experimental Studies of Morphological Stability 268

PARKER, R.L.

VII. Effects of Impurities on Crystal Growth Mechanisms 272
    25. Introduction 272
    26. Growth of Binary Chains 273
    27. Poisoning of Kinks during Layer Growth by Step Motion around Screw
        Dislocations 276
    28. Sucrose Growth from Solution: Step Pinning 278
    29. Step Pinning (Continued) 280
    30. Kinematic Waves and Impurities in Dissolution: Etch Pits 281
    31. Macroscopic Particles as the Impurity 283
VIII. Fluid Flow Effects 284
    32. Introduction 284
    33. Flow past a Flat Plate 286
    34. The Rotating Disk and the Czochralski Apparatus 290
    35. The Sphere in Fluid Flow 294
    36. Thermal or Natural Convection and Crystal Growth in Horizontal Boats. 295
    37. Fluid Flow Resulting from Crystal/Liquid Density Differences 298

PEERCY, P.S.: *see* SAMARA, G.A.

PEETERS, F.M.
VOLUME 38, 1984

Theory of Polaron Mobility

F.M. PEETERS

Department of Physics, University of Antwerp, Antwerp, Belgum

AND

J.T. DEVREESE

Department of Physics and Institute for Applied Mathematics,
University of Antwerp, Antwerp, Belgium
Department of Physics, Eindhoven University of Technology, Einhdhoven, The Netherlands

  I. Introduction 82
 II. A Boltzmann Equation for the Polaron 85
    1. The Fröhlich Polaron Hamiltonian 85
    2. The Feynman Polaron Model Hamiltonian 86
    3. The Polaron Boltzmann Equation for Arbitrary
       Electron–Phonon Coupling Strength 87
    4. First Moment of the Boltzmann Equation 90

PEETERS, F.M.

III. Relaxation-Time Approximation for Linear Conductivity.................... 91
    5. The Impedance Function for Small Frequencies ($\omega \ll \omega_{LO}$) 91
    6. Numerical Results for the Scattering Rate and the
       Impedance Function 94
IV. The Drifted Maxwellian Approach for Nonlinear Direct Current Conductivity.. 97
    7. Rederivation of the Thornber–Feynman Result 97
    8. Discussion of the Validity of the Thornber–Feynman Result
       and Numerical Analysis.. 100
V. The Polaron Impedance Function for Quantum Frequencies 107
    9. Rederivation of the Feynman–Hellwarth–Iddings–Platzman ....
       Result 107
   10. Discussion of the Validity of the FHIP Result
       and Numerical Analysis.. 109
VI. The Polaron Mass and the Polaron Mobility............................. 117
   11. Physical Picture of a Polaron 117
   12. The Temperature Dependence of the Polaron Mass 118
   13. The Polaron Mobility: The $\frac{3}{2}kT/\hbar\omega_{LO}$ Problem
       at Low Temperature... 123
VII. Conclusion 127
    Appendix A 129
    Appendix B... 130
    Appendix C... 132

PEREZ-ALBUERNE, E.A.: *see* DRICKAMER, H.G.

PETERSON, N.L.   VOLUME 22, 1968

Diffusion in Metals

N.L. PETERSON

Argonne National Laboratory, Argonne, Illinois

I. Preface... 409
II. Introduction... 410
   1. Mechanisms of Diffusion... 410
   2. Defect Formation and Motion..................................... 412
   3. Diffusion Coefficients.. 416
   4. Temperature and Pressure Dependence of Diffusion.................. 417
III. Self-Diffusion... 420
   5. Random Walk.. 420
   6. Correlation Effects.. 422
   7. Experimental... 428

PETERSON, N.L.

   IV. Impurity Diffusion.. 433
        8. Correlation Effects... 433
        9. Relation between Impurity and Self-Diffusion...................... 436
       10. Ultra-Fast Impurity Diffusion................................. 444
    V. Effect of Solute Additions on Self-Diffusion........................... 446
   VI. Tracer Diffusion in Concentrated Binary Alloys........................ 450
      11. Random Alloys.. 451
      12. Ordered Alloys... 453
  VII. Isotope Effects in Diffusion...................................... 457
 VIII. Diffusion in bcc Metals... 467
      13. Normal bcc Metals.. 468
      14. Anomalous bcc Metals....................................... 470
   IX. Diffusion in an Electrical Gradient................................. 481
      15. Electrical Gradients and Vacancy Flow.......................... 482
      16. The Force on the Jumping Ion................................. 484
      17. Experimental.. 486
    X. Position-Dependent Diffusion Coefficients........................... 491
      18. Diffusion in a Chemical Concentration Gradient.................... 491
      19. Diffusion in a Temperature Gradient............................ 499
   XI. Concluding Remarks.. 512

PETTIFOR, D.G.   VOLUME 40, 1987

A Quantum-Mechanical Critique of the Miedema Rules for Alloy Formation

D.G. PETTIFOR

*Department of Mathematics,
Imperial College,
London, England*

   I. The Miedema Scheme   43
     1. Introduction   43
     2. The Miedema Scheme   45
     3. The Ionic Model   50
  II. The Nature of the Metallic Bond   53
     4. Bonding in Simple Metals   53
     5. Simple Metal Heats of Formation   62
     6. Bonding in Transition Metals   67
     7. Transition Metal Heats of Formation   74
 III. Significance of the Miedema Parameters   87
     8. Significance of the Miedema Parameters and Conclusion   87

PFA

PFANN, W.G.   VOLUME 4, 1957

Techniques of Zone Melting and Crystal Growing

W.G. PFANN

Bell Telephone Laboratories, Murray Hill, New Jersey

I. Introduction... 424
II. The Distribution Coefficient and Normal Freezing.................... 426
   1. Equilibrium Freezing... 427
   2. Segregation during Normal Freezing............................. 429
   3. Burton-Prim-Slichter Theory for the Effective Distribution Coefficient 431
   4. Determining the Distribution Coefficient....................... 433
   5. General Remarks on Conditions of Crystallization................ 435
III. Principles of Zone Refining...................................... 437
   6. Definition of Terms; Fields of Application of Zone Melting........... 437
   7. Single-Pass Zone Melting....................................... 440
   8. Multi-Pass Zone Melting.. 443
   9. Calculation of Multi-Pass Distributions........................ 444
   10. The Ultimate Distribution...................................... 450
   11. Matter Transport.. 451
   12. Solid-Phase Zone Melting....................................... 452
   13. Continuous Zone Refining....................................... 452
IV. Practice of Zone Refining.. 456
   14. General Considerations... 456
   15. Zone Refining with a Container................................. 457
   16. Zone Refining without a Container.............................. 459
   17. Some Applications of Zone Refining............................. 463
      a. Germanium.. 463
      b. Silicon.. 464
      c. Other Semiconductors..................................... 464
      d. Aluminum... 465
      e. Iron... 465
      f. Zirconium.. 466
      g. Gallium.. 466
      h. Bismuth.. 467
      i. Antimony... 467
      j. Tin.. 467
      k. Chemicals.. 468
V. Zone Leveling.. 468
   18. Simple Zone Melting... 469
   19. Simple Zone Melting with Starting Charge....................... 469
   20. Starting Charge into Pure Solvent.............................. 470
   21. Repeated Pass... 470

PFANN, W.G.

VI. Methods of Perturbing the Concentration............................ 472
  22. Perturbations by Changing the Composition of the Melt............. 472
  23. Perturbations by Changing the Volume of the Melt................. 475
     a. Segregation Junctions.. 476
     b. Remelt Method... 477
  24. Perturbations by Changing the Growth Conditions................. 481
     a. Diffusion-Dominated Normal Freezing......................... 482
     b. Programmed Growth Rate in Normal Freezing................... 483
     c. Analogy between Enriched Film and Molten Zone............... 484
     d. p-n Junctions by Variation of Growth Conditions............. 485
  25. Diffusion Techniques... 492
VII. Temperature-Gradient Zone Melting................................. 493
  26. Principles.. 493
  27. Applications.. 495
     a. Semiconductive Devices...................................... 495
     b. Single Crystals... 499
     c. Joining... 499
VIII. Crystal Growth and Dislocations.................................. 499
  28. Introduction.. 499
  29. Fundamentals of Crystal Growth from the Melt.................... 501
     a. Heat Flow... 501
     b. Nucleation.. 502
     c. Orientation... 503
     d. Constitutional Supercooling................................. 505
  30. Techniques of Growth from the Melt.............................. 506
     a. Crystal Pulling... 506
     b. Zone Leveling... 509
  31. Dislocations in Crystals Grown from the Melt.................... 513
  32. Measurement of Crystal Perfection............................... 519

PHI

PHILLIPS, J.C.   VOLUME 18, 1966

The Fundamental Optical Spectra of Solids*

J.C. PHILLIPS

Department of Physics and Institute for the Study of Metals
The University of Chicago, Chicago, Illinois

I. Introduction... 56
  1. Scope of Review.. 56
  2. The One-Electron Approximation.................................... 57
  3. Electron–Phonon and Electron–Electron Interactions (Perturbative
     Approach)... 59
  4. Measurement of Optical Constants.................................. 61
  5. Surface Treatment and Low Temperatures............................ 63
  6. Anisotropic Crystals.. 64
  7. Magnetooptic Effects.. 64
  8. Plasma Oscillations... 66
II. One-Electron Spectra... 67
  9. Line Shape of Interband Transitions............................... 67
  10. Oscillator Strengths... 70
  11. Survey of Theoretical Energy Bands.............................. 73
  12. Spin-Orbit Splittings of Band Edges............................. 75
  13. Homopolar Semiconductors.. 76
  14. Zincblende Crystals... 90
  15. d Bands in Semiconductors..................................... 98
  16. Effective Masses and g-Factors................................ 100
  17. Noncubic Semiconductors... 101
  18. Alkali Halides.. 102
  19. Solid Rare Gases.. 110
  20. Diatomic V and IV–VI Crystals................................... 111
  21. Polyvalent Metals... 115
  22. Noble Metals.. 119
  23. Transition Metals... 123
III. Resonances... 125
  24. Excitons.. 125
  25. Parabolic Excitons.. 126
  26. Hyperbolic (Saddle-Point) Excitons.............................. 128
  27. Resonance Behavior in Many-Body Systems......................... 131
  28. Examples of Asymmetric Isolated Resonances...................... 136

PHILLIPS, J.C.

29.	Inhomogeneous Broadening	139
30.	Exciton Families in Xe	141
31.	Alkali Halides	144
32.	Semiconductors	146
33.	Hybrid Excitons	150
34.	Strong Exciton–Phonon Interactions	152
35.	Alkali Metals	155
IV.	Summary	160
V.	Recent Developments	161

PHILLIPS, J.C.   VOLUME 37, 1982

Spectroscopic and Morphological Structure of Tetrahedral Oxide Glasses

J.C. PHILLIPS

Bell Laboratories, Murray Hill, New Jersey

I.	Introduction	93
II.	Schematic Spectrum of a Tetrahedral Oxide Glass	97
III.	Neutron Scattering Spectra: g-SiO$_2$	100
IV.	Infrared Spectra: g-SiO$_2$	103
V.	Raman Spectra: g-SiO$_2$	106
VI.	Dispersion of Internal Surface Modes	108
VII.	Cluster Dimensions and Morphology (SiO$_2$)	115
VIII.	Morphology of Clusters in g-GeO$_2$	121
IX.	Ring Statistics and Dynamical Force Field Constraints	124
X.	Vibrational Spectroscopy of g-GeO$_2$	128
XI.	Electronic Structure	133
XII.	Structure and Vibrational Spectra of Alkali Silicate Glasses	137
XIII.	Structure and Vibrational Spectra of Alkali Germanate Glasses	146
XIV.	Ternary Alkali Germanate–Silicate Glasses	149
XV.	Pb Silicate and Germanate Glasses	151
XVI.	BeF$_2$: A Nonoxide Tetrahedral Glass	153
XVII.	Spectroscopic Analogs of Low-Temperature Thermodynamic Anomalies	156
XVIII.	High-Resolution Electron Micrographs	158
XIX.	Phase Separation in Silicates	161
XX.	Frustration of the Crystallization Path	162
XXI.	Conclusions	168
	Appendix A. Combined AX$_4$ and A$_2$X Symmetries	168
	Appendix B. Clusters and Small-Angle X-Ray Scattering	170

PHI–PIC

PHILLIPS, T.G.:   *see* HENSEL, J.C.

PICKETT, Warren E.   VOLUME 48, 1994

Electrons and Phonons
in C_{60}-Based Materials

WARREN E. PICKETT

Complex Systems Theory Branch
Naval Research Laboratory
Washington, D.C.

I.	Introduction	226
1.	Background	226
2.	Generalities	229
II.	Electrons on the C_{60} Molecules	231
3.	Structure and Symmetry	231
4.	Electronic Structure	234
III.	Dynamics of the C_{60} Molecule and Solids	253
5.	Symmetry and the Molecular Vibrations	253
6.	Fullerite C_{60}	258
7.	Fullerides $A_n C_{60}$	269
8.	Theory of Jahn–Teller Distortions	272
IV.	Crystal Structures	276
9.	Fullerite C_{60}	276
10.	Alkali Fullerides	283
11.	Alkaline-Earth Fullerides; Others	286
V.	C_{60} Crystals: Local Density Calculations	289
12.	Fullerite	290
13.	Alkaline Fullerides	296
14.	Alkaline-Earth Fullerides; Others	301
15.	Tight-Binding Representation	303
VI.	Electronic Spectroscopies; Coulomb Interactions	305
16.	Photoemission	305
17.	Bulk Spectroscopies	310
18.	Correlation; Magnetic Behavior	316
VII.	Electron–Phonon Coupling	318
19.	Length and Energy Scales	318
20.	Formal Theory	319
21.	Theoretical Estimates of Coupling Strength	321
VIII.	Transport, Thermodynamic, and Infrared Properties	323
22.	d.c. Transport Data	324
23.	Heat Capacity; Infrared Response	327
24.	Theoretical Treatments of Transport	328

PICKETT, Warren E.

IX. Superconductivity 332
   25. Properties of the Superconducting State 332
   26. Isotope Effect 339
   27. Coulomb Pseudopotential μ^* 341
   28. Proposed Mechanisms 342
 X. Summary 346
XI. Acknowledgments 346

PIERANSKI, P.:   *see* DUBOIS-VIOLETTE, E.

PINES, David   VOLUME 1, 1955

Electron Interaction in Metals

DAVID PINES

Palmer Physical Laboratory, Princeton University, Princeton, New Jersey

  I. Introduction.. 368
   1. Survey.. 368
   2. Independent Electron Model.................................. 371
   3. Previous Treatments of Electron Interaction................. 374
 II. The Collective Description of Electron Interactions................... 376
   4. Classical Theory and Physical Picture....................... 376
   5. The Canonical Transformation to the Collective Representation........ 382
     a. Introduction of Plasma Degrees of Freedom.................... 382
     b. The Basic Collective Hamiltonian............................ 385
     c. The Transformation to Independent Collective Modes............. 388
   6. Correlation Energy of the Free Electron Gas................. 391
     a. First Approximation: the Choice of β................. 391
     b. The Total Correlation Energy............................... 395
   7. The Influence of the Lattice............................... 400
     a. The Frequency of Plasma Oscillation......................... 400
     b. The Correlation Energy..................................... 405
III. The Influence of Electron Interactions on Various Metallic Properties..... 407
   8. Nonmagnetic Metallic Properties............................ 407
     a. Level Density and Specific Heat............................ 407
     b. Soft X-ray Emission Spectra and the Conduction Band Width...... 409
     c. Transport Properties of Metals............................. 412
     d. Electron-Electron Scattering in Metals..................... 414
   9. Magnetic Properties.. 416
     a. Spin Paramagnetism... 416
     b. Nuclear Magnetic Resonance in Metals....................... 420
     c. Orbital Electron Diamagnetism.............................. 424
     d. Ferromagnetism... 427

PIN–PIP

PINES, David

IV. The Excitation of Plasma Oscillations in Solids...................... 429
  10. Characteristic Energy Loss Spectra.............................. 429
    a. Theoretical Consideration..................................... 429
    b. Experimental Loss Values..................................... 433
    c. Angular Distribution of Characteristic Energy Losses............ 437
    d. Stopping Power of the Conduction Electrons.................... 440
  V. The Electron-Lattice Interaction................................... 441
  11. The Role of Electron Interaction............................... 441
  12. The Collective Description of Coupled Electron-Ion Motion.......... 446
VI. Conclusion... 449
  13. The Relationship of Electron Interaction to Other Many-Body Problems 449

PIPER, W.W.   VOLUME 6, 1958

Electroluminescence

W.W. PIPER AND F.E. WILLIAMS

General Electric Research Laboratory, Schenectady, New York

  I. Introduction.. 96
 II. Background Theory.. 98
  1. Insulators and Semiconductors.................................. 98
  2. Dielectric Breakdown.. 102
  3. Luminescence... 105
III. Basic Mechanisms of Electroluminescence........................... 113
  4. Excitation Processes.. 114
    a. Injection of Minority Carriers................................ 114
    b. Valence Band Ionization..................................... 117
    c. Direct Field Ionization of Impurities.......................... 118
    d. Acceleration of Charged Carriers to Optical Energies............ 120
  5. Energy Transfer Processes...................................... 125
    a. Transport and Capture of Charge Carriers...................... 125
    b. Collision Excitation... 126
    c. Exciton and Resonance Transfer Processes...................... 127
  6. Emission.. 128
    a. Impurity Systems... 128
    b. Interband Radiation... 133
    c. Avalanche Radiation... 137

PIPER, W.W.

PLA

PLATZMAN, P.M.   SUPPLEMENT 13, 1973

Waves and Interactions in Solid State Plasmas

P.M. PLATZMAN

Bell Telephone Laboratories, Murray Hill, New Jersey

P.A. WOLFF

Department of Physics, Massachusetts Institute of Technology, Cambridge, Massachusetts

PREFACE.   ix

I. Characteristics of Solid State Plasmas

1. What Is a Plasma ?   1
2. Conditions for Weak Coupling   3
3. Fermi Liquid Theory—Introductory Remarks   5
4. Coulomb Energy in a Plasma   7
5. What Parameters Characterize a Plasma ?—Screening Lengths   8
6. The Characteristic Frequency   11
7. Other Plasma Parameters   15
    References   19

II. Theory of Important Plasma Experiments

8. Qualitative Discussion of Scattering Experiments   20
9. Analysis of Electron Scattering Experiments   23
10. Scattering from a Noninteracting Electron System   26
11. Light Scattering—The Classical, Single-Component Case .   27
12. Transmission Experiments: Introductory Remarks   31
13. Response Functions   33
14. Response to the Total Field   37
15. The Plasma Dielectric Constant   40
16. Response to Transverse Fields   43
17. The Fluctuation–Dissipation Theorem   47
    References   48

III. The Random Phase Approximation

18. The Self-Consistent Field Method   49
19. Discussion of the Random Phase Approximation   52
20. Static Properties of a Degenerate Plasma   55
21. High-Frequency Properties of a Degenerate Plasma   57
22. Screening in Classical Plasmas   59
23. Plasma Waves in Maxwellian Plasmas   60
    References   62

PLATZMAN, P.M.

IV. Comparison with Experiment—Degenerate Plasmas

24. RPA Formula for the Spectrum 64
25. Integrated Spectrum for Characteristic Energy Loss
    Experiments 65
26. Mean Free Path For Plasmon Emission 66
27. Surface Plasmons 67
28. Angular Variation of Characteristic Energy Loss Spectra . . 68
29. Exhange Corrections—The Generalized Random Phase
    Approximation 71
30. The Cutoff Angle and Decay of Long-Wavelength Plasmons. 75
31. Effects of Band Structure 77
32. Plasmons in Doped Semiconductors 80
    References 81

V. Light Scattering Experiments

33. Introduction 82
34. Plasmon Light Scattering and Plasmon–Phonon Coupling . 85
35. The Magnetoplasma and Light Scattering Experiments
    Involving Magnetized Plasmas 90
36. Multicomponent Plasmas 99
37. Light Scattering by Multicomponent Plasmas 107
38. Band Structure Effects in Light Scattering 111
    References 117

VI. Waves in Metals

39. Introductory Remarks 119
40. General Formulation of the Wave Propagation Problem . . 123
    References 125

VII. The Local Limit

41. The Dispersion Relation for Helicons 127
42. The Dielectric Tensor 131
43. Alfven Waves 142
44. Helicon–Phonon Interaction 144
    References 150

VIII. Nonlocal Effects in a Noninteracting Gas

45. The Solution of the Boltzmann–Vlasov Equation 151
46. Helicons in the Nonlocal Regime 157

PLA

PLATZMAN, P.M.

47. Helicon Propagation at an Angle to the Magnetic Field . . 161
48. Cyclotron Wave Propagation in a Noninteracting Gas . . . 164
49. The "Problem" of Boundaries 191
    References 202

IX. Conduction Electron Spin Resonance in an Electron Gas

50. Hydrodynamic or Long-Wavelength Treatment of CESR . 204
51. CESR in a Bounded Medium 212
    References 219

X. The Interacting Electron Liquid

52. The Phenomenological Formulation of Fermi-Liquid
    Theory 220
53. Equilibrium Properties of the Interacting Electron Liquid . 227
54. Transport Equations for the Interacting Fermi Liquid . . 232
55. The Electrical Transport Equation for Fermi Liquids . . . 241
56. CW in the Presence of Interactions 245
57. CW Propagating Parallel to the Magnetic Field 252
58. Spin Waves in the Interacting Electron Liquid 256
    References 271

XI. The Microscopic Theory of Fermi Liquids

59. Introduction to the Microscopic Theory of Fermi Liquids . 273
60. The General Formulation of the Microscopic Theory . . . 274
61. Coulomb Interactions and the Landau Scattering Function . 283
62. The Electron–Phonon Interaction and the Landau
    Scattering Function 287
63. Comparison of Numerical Estimates and Experimental
    Results 290
    References 294

ADDENDUM: ALKALI METAL PARAMETERS 295
AUTHOR INDEX 297
SUBJECT INDEX 301

POIRIER, D.M.: *see* WEAVER, J.H.

POND, R.C. VOLUME 47, 1994

Defects at Surfaces and Interfaces

R.C. POND

*Department of Materials Science and Engineering, University of Liverpool,
Liverpool, United Kingdom*

J.P. HIRTH

*Department of Mechanical and Materials Engineering, Washington State University,
Pullman, Washington*

I.	Introduction	288
II.	Crystallographic Tools and Methodology	293
	1. The Structure of Crystalline Objects	294
	2. The Symmetry of Infinite Crystals	295
	3. The Symmetry of Semi-Infinite Crystals	299
	4. The Symmetry of Bicrystals	300
	5. Defect Characterization: The Class of Manifest Symmetry	304
	6. Defect Characterization: The Class of Broken Symmetry	309
	7. Summary	313
III.	Defects in Single Crystals	315
	8. Line Defects: The Class of Manifest Symmetry	315
	9. Extended Defects: The Class of Broken Symmetry	318
	10. Imperfect Defects	320
IV.	Surface Discontinuities	321
	11. Defects Due to Broken Symmetry on Unrelaxed Surfaces	322
	12. Defects Due to Broken Symmetry on Reconstructed Surfaces	324
	13. Defects on (001) Surfaces of Silicon Crystals	325
	14. Conservation of Topological Properties	327
V.	Interfacial Discontinuities	330
	15. *A Priori* Characterization	330
	16. *A Posteriori* Characterization	333
	17. Interfacial Defects in the Class of Broken Symmetry: Surface-Like Discontinuities	334
	18. Interfacial Dislocations in the Class of Broken Symmetry	338
	19. Other Interfacial Defects	349
VI.	Interfaces Modeled as Dislocation Arrays	350
	20. Mapping Into a Single Reference Space: The Frank-Bilby Equation	351
	21. Mapping Into a Dichromatic Complex	356
	22. Dislocation Content of Epitaxial Interfaces	359
	Acknowledgments	365

RAB–RAS

RABIN, Herbert: *see* COMPTON, W. Dale

RAJAGOPAL, A.K.:   *see* JOSHI, S.K.

RASOLT, M.   VOLUME 43, 1990

Continuous Symmetries and Broken Symmetries in Multivalley Semiconductors and Semimetals

M. RASOLT

Solid State Division
Oak Ridge National Laboratory
Oak Ridge, Tennessee

I.	Introduction	94
II.	Isospin Space and the Effective-Mass Hamiltonian	95
	1. Band-Structure Symmetry	95
	2. Isospin and the Effective-Mass Hamiltonian	98
	3. Ferromagnetic Alignment of the Isospin–Spin Ground State	102
	4. Antiferromagnetic Alignment in Isospin–Spin Space	105
	5. Other Broken Symmetries in Isospin–Spin Space	109
	6. Isospin–Spin Symmetry in the Presence of a Uniform Magnetic Field	113
III.	Symmetry-Breaking Terms in the Effective-Mass Hamiltonian	114
	7. The Energy of $\hat{H}$ in the RPA	114
	8. Effective Hamiltonian for the Valley Electrons	116
	9. The Effective-Mass Hamiltonian	122
	10. Symmetry-Breaking Terms	125
	11. Effect of an External Magnetic Field	130
IV.	Isospin Space and the Integer Quantum Hall Effect	132
	12. Isospin Polarized Ground State in the (110) Silicon Inversion Layer	133
	13. Effect of Symmetry-Breaking Terms on the Nature of the Isospin Polarization	138
	14. Effect of Disorder on the Nature of the Isospin Polarization	142
	15. Valley Waves in the 110 Inversion Layer of Si	148
	16. Dissipation in Isospin Space	151

RASOLT, M.

V. Isospin Space and the Fractional Quantum Hall Effect 156
   17. The Fractional QHE (FQHE) Ground State in Isospin
      Space 156
   18. The Single Mode Approximation (SMA) 165
   19. Collective Modes in the Fully Isospin Polarized FQHE 168
   20. Collective Modes in the Isospin Unpolarized FQHE 175
   21. Isospin Representation of the QHE in a Two-Layer Heterojunction..... 178
VI. Isospin in Three Dimensions 182
   22. Isospin BS in Three Dimensions in the Absence of an
      External Magnetic Field 183
   23. Isospin BS in Three Dimensions in the Presence of an
      External Magnetic Field 187
   24. Superconductivity in a Strong Magnetic Field........................ 194
   25. Effect of Quantum Fluctuations.................................... 209
   26. Isospin Fluctuations in the Fermi Liquid State 214
   Acknowledgments 223
   Appendix A 223
   Appendix B 226

REIF, F.: *see* COHEN, M.H.

REITZ, John R. VOLUME 1, 1955

Methods of the One-Electron Theory of Solids

JOHN R. REITZ

Case Institute of Technology, Cleveland, Ohio

  I. Introduction.. 2
 II. The Fock Equation... 3
   1. Derivation of the Fock Equation................................. 4
   2. The Simpler Hartree Equation.................................. 11
   3. Localized Solutions and Crystal Orbitals........................ 12
   4. Correlated Motions of Electrons................................. 14
III. Historical Survey of the One-Electron Method...................... 15
 IV. General Properties of Crystal Orbitals.............................. 18
   5. Bloch Waves.. 19
   6. Some One-Dimensional Examples................................ 20
   7. Further Properties of Bloch Waves............................... 23

REI

REITZ, John R.

V. Brillouin Zones.. 26
    8. Brillouin Zones: the Result of Crystal Symmetry................... 26
    9. Brillouin Zones: from Perturbation Theory........................ 29
    10. The Brillouin Zone and the Jones Zone........................... 32
    11. Examples of Zone Structure...................................... 34
        a. Face-Centered Cubic Lattice................................... 34
        b. Body-Centered Cubic Lattice................................... 36
        c. Hexagonal Lattice... 37
        d. The Diamond Structure... 38
VI. The Mathieu Problem and Energy Bands in Crystals................... 38
    12. The One-Dimensional Mathieu Problem............................. 39
    13. Momentum Eigenfunctions for the One-Dimensional Mathieu Equation 41
    14. The Two-Dimensional Mathieu Problem............................. 44
VII. The Tight Binding Method.. 46
    15. Conventional Treatment of Tight Binding......................... 47
        a. Simple Cubic Lattice—s State................................ 48
        b. Body-Centered Cubic Lattice—s State......................... 49
        c. Face-Centered Cubic Lattice—s State......................... 50
        d. Hexagonal Close-Packed Structure—s State.................... 50
        e. Simple Cubic Lattice—p State................................ 51
    16. Extensions of the Tight Binding Method.......................... 52
    17. Orthogonalized Atomic Orbitals................................. 54
    18. LCAO Method as an Interpolation Procedure....................... 58
VIII. The Cellular Method.. 61
    19. The Wigner-Seitz Approximation.................................. 62
        a. The Quantum Defect Method..................................... 66
    20. Expansion of the Wave Function in Spherical Harmonics. Lattice
        Harmonics... 66
        a. Lattice Harmonics... 72
    21. The Empty Lattice Test.. 72
    22. Application of the Cellular Method to Composite Lattices......... 73
IX. Variational Methods for Periodic Lattices......................... 75
    23. Variational Principle for Periodic Lattices..................... 76
X. The Orthogonalized Plane-Wave Method............................... 79
    24. Expansion of the Wave Function in Orthogonalized Plane Waves..... 80
    25. Difficulties with the Orthogonalized Plane-Wave Method.......... 83
XI. Other Crystal Orbital Methods..................................... 84
    26. Augmented Plane-Wave Method..................................... 86
XII. The Heitler-London Method... 88
    27. The Landshoff-Löwdin Method for Ionic Crystals.................. 88
    28. Localized Bond Orbitals for Diamond............................. 91
XIII. Exchange Interaction in the One-Electron Approximation............ 92
    29. The Exchange Energy of Free Electrons........................... 92
    30. The Exchange Energy of Tightly Bound Electrons.................. 94

Compression of Solids by Strong Shock Waves

M.H. RICE, R.G. MCQUEEN, AND J.M. WALSH

Los Alamos Scientific Laboratory, Los Alamos, New Mexico

I. Introduction.. 1
II. Basic Relations from Fluid Dynamics................................. 3
   1. The Conservation Relations for Continuous Flow..................... 3
   2. The Flow in a Simple Rarefaction Wave............................. 4
   3. The Conservation Relations for a Shock Wave....................... 7
   4. The Stability of Shock Waves...................................... 9
      a. Two-Wave Structures Arising from Material Rigidity............. 10
      b. Two-Wave Structures Arising from a Phase Change............... 12
III. Experimentation... 15
   5. Producing Strong Shock Waves...................................... 15
   6. The Measurement of Shock Wave Parameters.......................... 17
      a. Pin-Contactor Experiments..................................... 17
      b. Photographic Experiments...................................... 20
      c. Other Experimental Approaches................................. 28
   7. The Transformation of Experimental Velocities to Pressure-Compression
      Points... 29
      a. Free-Surface Velocity Approximation........................... 29
      b. Impedance Matching with 24ST Aluminum......................... 30
   8. Discussion of Experimental Results................................ 33
IV. Equation of State... 40
   9. Introduction.. 40
  10. The Mie-Grüneisen Equation of State.............................. 41
  11. The Grüneisen Ratio as a Function of Volume...................... 43
      a. The Slater Relation for a Continuum........................... 43
      b. The Dugdale-MacDonald Relation................................ 44
      c. Barron's Calculations... 46
      d. Uncertainties in the General Equation of State................ 46
  12. The Thermodynamic Description of High-Pressure States............ 47
V. A Test of the Dugdale-MacDonald Formula............................. 60

RICE, T.M.:   *see* HALPERIN, B.I.

RIC–RIT

RICE, T.M.   VOLUME 32, 1977

The Electron–Hole Liquid in Semiconductors: Theoretical Aspects

T.M. RICE

Bell Laboratories, Murray Hill, New Jersey

I. Introduction	1
II. Semiconductor Systems	5
1. Band Structure	5
2. Excitons	10
III. Ground-State Energy of Electron–Hole Systems	13
3. Ground State of Exciton Molecules; Molecular Liquid?	13
4. Metallic Electron–Hole Liquid	20
5. Single-Particle Properties	35
6. Conclusions	38
IV. General Properties of a Metallic Condensate	39
7. The Phase Diagram of the Electron–Hole Fluid	39
8. The Luminescence from the Electron–Hole Liquid	50
9. Theory of Electromagnetic Effects	56
10. Magnetic Field Effects	63
11. Impurity Effects	66
V. Theory of Finite Size Drops, Their Growth and Death	70
12. Surface of an Electron–Hole Drop	70
13. Charge Structure of a Drop	74
14. Rate Equations	76
15. Nucleation and Size of Drops	81

RITCHIE, R.H.: *see* ECHENIQUE, P.M.

ROITBURD, A.L.   volume 33, 1978

Martensitic Transformation as a Typical Phase Transformation in Solids

A.L. ROITBURD

Institute of Metal Physics and Metallography, IP Bardin Central Scientific Research Institute of Ferrous Metallurgy, Moscow, USSR

I.   Introduction 317
II.   Initial Theses 327
      1. Strain Phase Transition 327
      2. Coherent State of Heterophase System 331
III. Equilibrium Structure and Thermodynamics 334
      3. Internal Stresses and Elastic Energy of Heterophase System 334
      4. Equilibrium of Interphase Boundaries 337
      5. Monodomain Plate of New Phase 339
      6. Polydomain Plates 348
      7. Elastic Domains and ''Phase Rule'' for Polydomain Structures 361
      8. Interaction between Plates and Ensembles of Crystals; Hierarchy of
         Structures 365
      9. Martensitic and Nonmartensitic Transformations 371
IV.   Kinetics of Martensitic Transformations 374
      10. Kinetics of Strain Transition, Coherent Interphase Boundary Energy and
          Mobility 374
      11. Noncoherent Boundary Mobility and Self-Diffusion Kinetics of Transfor-
          mation 381
      12. Crystal Nucleation, Structure-Kinetic Types of Transformations 385
V.   Conclusion 389

ROSENBERG, H.M.:   *see* MENDELSSOHN, K.

SAF–SAI

SAFRAN, S.A. VOLUME 40, 1987

Stage Ordering in Intercalation Compounds

S.A. SAFRAN

Exxon Research and Engineering,
Corporate Research Science Laboratories,
Annandale, New Jersey

I. Staging: Modulated Structure and Effects on Properties	183
1. Introduction	183
2. Intercalation Processes and Staging	186
3. Staging Effects on Host Properties	193
II. Mechanisms of Stage Ordering	206
4. In-Plane Interactions	206
5. Interlayer Interactions	210
III. Staging Transitions	214
6. Phase Diagrams for Staging Transitions	215
7. Pressure-Induced Staging Transitions	226
IV. Domains and Staging Kinetics	233
8. Domain-Induced Random Staging	233
9. Domain–Domain Interactions	238
10. Kinetics of Staging and Intercalation	241

SAITO, N. VOLUME 14, 1963

Molecular Motion in Solid State Polymers

N. SAITO

Department of Applied Physics, Waseda University, Tokyo, Japan

AND

K. OKANO, S. IWAYANAGI, AND T. HIDESHIMA

The Institute of Physical and Chemical Research, Tokyo, Japan

I. Introduction	344
II. Flexibility of a Polymer	346
1. Internal Rotation	346
2. Stereochemical Structure and Potential of Hindered Rotation	347

SAITO, N.

III. Statistical Mechanics of the Equilibrium Conformation of Polymers 349
    3. General Considerations 349
    4. Effect of the Hindered Rotation 353
    5. Excluded Volume Effect 360

IV. Elastic Properties of Polymers 363
    6. Elasticity of a Single Polymer Chain 363
    7. Elasticity of Gaussian Networks 365
    8. Large Deformation 367

V. Statistical Mechanics of Nonequilibrium Conformation of Polymers 369
    9. Distribution Function of the Conformation of Flexible Polymers 369
    10. Rouse Model 371

VI. Viscoelastic Properties of Amorphous Polymers 374
    11. Phenomenological Theory of Linear Responses to the Stimuli 374
    12. Molecular Theory of Viscoelasticity of Amorphous Polymers 378

VII. Classification of Dispersion and Absorption in Polymers 387
    13. Classification of the Modes of Motions of Backbone Chain 387
    14. Molecular Motions at High Temperatures 389
    15. Molecular Motions at Low Temperatures 391

VIII. Dispersion Associated with the Micro-Brownian Motion of the Main Chains 398
    16. Viscoelastic and Dielectric Relaxations and Nuclear Magnetic Resonance 398
    17. Measurements of Volume Viscoelasticity 405
    18. Theory of Volume Viscoelasticity 407
    19. Identification of the Primary Dispersion in Semicrystalline Polymers . . . 412
    20. Relaxation Spectrometry of Semicrystalline Polymers 416

IX. Dispersion Associated with the Local Relaxation Mode 419
    21. Introductory Remarks 419
    22. Observations Concerning the Local Relaxation Mode 425

X. Dispersion Associated with Motion of Side Chains 432
    23. General Survey of Side Chain Relaxation 432
    24. Molecular Motions in Polymethyl Methacrylate 435
    25. Relaxation Spectrometry of the Side Chain Relaxation in Poly-n-Alkyl
        Methacrylates 441

XI. Molecular Motions in Semicrystalline Polymers 453
    26. α-Dispersion in Semicrystalline Polymers 453
    27. Semicrystalline Polymer as a Mixture 456
    28. Molecular Motions in Polymer Crystals 463

XII. Glass Transition in Polymers 468
    29. Introductory Remarks 468
    30. The Glass-Transition Phenomenon 469
    31. Origin of the Glass Transition 473
    32. Relaxational Feature of the Glass Transition 475
    33. Temperature Dependence of the Relaxation Process 478
    34. Freezing-in of the Free Volume and the Glass Transition 485
    35. Manifestations of the Glass Transition 494
    36. Concluding Remarks to Part XII 501
    Acknowledgments 502

SAMARA, G.A.   VOLUME 36, 1981

The Study of Soft-Mode Transitions at High Pressure*

G.A. SAMARA AND P.S. PEERCY

Sandia National Laboratories, Albuquerque, New Mexico

I. Introduction 2
II. Theoretical Background 5
   1. Conditions for Lattice Stability and the Soft-Mode Concept 5
   2. Landau's Theory for Continuous Phase Transitions 6
   3. Harmonic Model 10
   4. Anharmonic Interactions 13
   5. Pressure Effects 17
   6. The Lyddane–Sachs–Teller (LST) Relation 18
   7. Coupled Phonon Modes 19
III. Displacive Ferroelectric Transitions 22
   8. Ferroelectricity and Incipient Ferroelectricity in Simple
      Diatomic Crystals 22
   9. Ferroelectricity in the Perovskite Structure........................ 26
   10. Other Displacive Ferroelectrics 44
   11. The Grüneisen Parameter of the Soft Ferroelectric Mode 53
IV. Displacive Antiferroelectric and Antiferrodistortive Transitions:
    Soft Short-Wavelength Optic Phonons 56
   12. The Antiferroelectric Perovskites 57
   13. Antiferrodistortive Transitions in $SrTiO_3$ and Its Isomorphs 60
   14. Antiferrodistortive Transitions in $Gd_2(MoO_4)_3$ and $BaMnF_4$ 62
   15. Theoretical Treatment 63
V. Potassium Dihydrogen Phosphate and Its Isomorphs 69
   16. General Properties 69
   17. Theoretical Model for the Transition............................... 72
   18. Soft Mode in the Paraelectric Phase 75
   19. Soft Mode in the Ferroelectric Phase 81
   20. Pressure Dependence of the Dielectric Properties and the
       Vanishing of the Ferroelectric State at High Pressure................ 85
   21. Relation between Macroscopic Properties and Microscopic
       Parameters of the Coupled Mode Model 88
   22. Effects of Deuteration and Pressure on the Dynamic Properties
       of KH_2PO_4 90
   23. The Case of the Antiferroelectric Crystals......................... 94
VI. Other Soft-Mode Systems... 95
   24. Paratellurite—TeO_2 96
   25. Rare Earth Pentaphosphates 101
   26. β-Tungsten Superconductors.................................... 103
VII. Effects of Pressure on Coupled Phonon Modes 105
   27. Optic–Acoustic Mode Coupling: Piezoelectric Crystals 105
   28. Optic–Acoustic Mode Coupling: Nonpiezoelectric Crystals 110
   29. Optic–Optic Mode Coupling 111
VIII. Concluding Remarks 113

SAMARA, G.A.   VOLUME 38, 1984

High-Pressure Studies of Ionic Conductivity in Solids

G.A. SAMARA

Sandia National Laboratories, Albuquerque, New Mexico

I.	Introduction	1
II.	Theoretical Background	4
	1. Lattice Defects	4
	2. Temperature Dependence of the Conductivity	5
	3. Pressure Dependence of the Conductivity: Activation Volume	8
	4. Model Calculations of Lattice-Defect Properties	11
III.	Results and Discussion	20
	5. Alkali Halides	23
	6. The Silver Halides AgCl and AgBr	32
	7. The Thallous Halides TlCl, TlBr, and TlI	36
	8. Fluorites and Other RX_2-Type Halides	44
	9. Fast Ion Conductors	65
IV.	Concluding Remarks	75

SCANLON, W.W.   VOLUME 9, 1959

Polar Semiconductors

W.W. SCANLON

U.S. Naval Ordnance Laboratory, Silver Spring, Maryland

I.	Introduction	83
II.	Crystal Properties	84
	1. Cohesive Forces in Crystals	84
	2. Crystal Structure	85
	3. Lattice Energy	86
	4. Plasticity of Crystals	87

SCA–SCH

SCANLON, W.W.

III. Physical Chemical Properties...................................... 92
    5. Vacancies in a Polar Crystal Lattice............................ 92
    6. Phase Relations... 93
    7. Crystal Growing.. 94
    8. Diffusion.. 98
    9. Crystal-Vapor Equilibrium Relationships........................ 102
    10. Foreign Impurity Atoms in Polar Semiconductors................. 106
    11. Alloys of the Pure Compounds.................................. 108
IV. Electrical and Optical Properties................................. 109
    12. Band Structure... 109
    13. The Intrinsic Energy Gap...................................... 111
    14. Optical Constants.. 118
    15. Transport Phenomena.. 118
      a. Hall Effect and Resistivity................................. 119
      b. Mobility... 125
      c. Mobility in Alloys... 131
      d. Thermoelectric Power....................................... 131
      e. Thermal Conductivity....................................... 133
    16. Recombination and Trapping.................................... 134
V. Conclusions... 136
VI. Acknowledgments... 137

SCHAFROTH, M.R.   VOLUME 10, 1960

Theoretical Aspects of Superconductivity

M.R. Schafroth

School of Physics,
The University of Sydney, Australia

Introduction... 295

Phenomenology

I. Thermodynamics.. 297
    1. General Theory for Bulk Matter................................ 297
    2. Two-Fluid Models... 300
    3. Surface Energy... 306
II. The Static Meissner Effect...................................... 311
    4. General Relations.. 311
    5. The London Kernel.. 324
    6. Non-London Kernels... 337
    7. Nonlinear Effects.. 358
    8. The Magnetostatics of the Intermediate State.................. 363
III. General Electrodynamics of Superconductors..................... 366
    9. Generalities.. 366
    10. Time-Dependent Magnetism..................................... 375
    11. Persistent Currents.. 389

SCHAFROTH, M.R.

Microscopic Theory

IV. General Aspects.. 401
   12. Introduction... 401
   13. Generalities... 402
   14. The Perfect Bose Gas Model.................................. 422
   15. Energy Gap.. 430
V. Single-Electron Theories... 439
   16. Theories Which Do Not Include Electron-Lattice Interaction......... 439
   17. Phonon Interaction with Perturbation Theory..................... 445
VI. Pair-Correlation Theories... 454
   18. Assumptions of the Theory of Metals........................... 454
   19. Pair Correlations.. 456
   20. The Theory of Bardeen, Cooper, and Schrieffer.................. 465
   21. The Bogoljubov Theory...................................... 473
   Bibliography.. 488

SCHNATTERLY, S.E.   VOLUME 34, 1979

Inelastic Electron Scattering Spectroscopy

S.E. SCHNATTERLY

Joseph Henry Research Laboratories, Princeton University, Princeton, New Jersey

I. Introduction 275
II. Elastic Scattering 278
   1. Born Approximation and Wentzel Model 278
   2. Electron Diffraction and Radiation Damage 280
III. Inelastic Scattering.. 282
   3. Born Approximation Expression 282
   4. Wentzel Model and Total Cross Section 286
   5. Approximate Mean Free Paths..................................... 288
   6. Oscillator Strength Sum Rule 290
   7. Kramers–Kronig Relations and Multiple Scattering 292
   8. Localized States and Angular Momentum Selection Rules 294
   9. Local Field Effects 294
IV. Basic Spectrometer Description 296
   10. Energy Selectors 298
   11. Momentum Transfer Selection 303
   12. Detectors.. 304
V. Sample Preparation 305
   13. Thin Film Evaporation 306
   14. Sputtering 308
   15. Thinning Macroscopic Crystals 308
VI. Experimental Examples 309
   16. Plasmons in Simple Metals 309
   17. Core Spectra of Simple Metals................................... 329
   18. Semiconductors and Insulators: Valence Excitations 340

SCHULMAN, James H.:   *see* KLICK, Clifford, C.

SCHWARTZ, L.M.:   *see* ENRENREICH, H.

SEITZ, Frederick: *see* WIGNER, Eugene P.

SEITZ, Frederick   VOLUME 2, 1956

Displacement of Atoms during Irradiation

FREDERICK SEITZ AND J.S. KOEHLER

University of Illinois, Urbana, Illinois

I. 1. Introduction 307
   2. Behavior of the Solid Types 308
II. Basic Equations of Collision Theory for Coulomb Encounters 311
   3. Classical Energy Relations 311
   4. Rutherford Scattering 314
   5. Effect of Screening 315
      a. Weak Screening 316
      b. Strong Screening 317
      c. General Relations 317
      d. Limiting Energies 319
   6. Behavior of Knocked-On Atoms 319
   7. Cross Section for Incident Particle 321
   8. Cross Sections for Displaced Atoms; Displacement Spikes 323
      a. Estimate of Mean Free Path for Hard-Sphere Collision 323
      b. Brinkman's Evaluation of Energy Transfer 324
   9. Rate of Loss of Energy by Incident Atom from Coulomb Encounters . . 327
III. Coulomb Encounters in Relativistic Range 328
   10. Energy Relations 328
   11. Rutherford Scattering of Electrons 329
   12. Effect of Screening 334
   13. Application to Heavy Particles 335

SEITZ, Frederick

IV. Neutron Collisions.. 336
  14. Nuclear Encounters.. 336
V. Electron Excitation... 338
  15. Qualitative Principles.. 338
  16. Model of Excitation Process for Comparatively Slow Velocities....... 340
  17. Excitation of Conduction Electrons.............................. 343
  18. Equations for Energy Loss.. 347
  19. Ratio of Energy Lost in Coulomb Encounters to That in Electron
     Excitation... 350
  20. Approximate Range-Energy Relations............................ 350
VI. Theory of Temperature Spikes.. 351
  21. Spherical and Cylindrical Spikes.................................. 351
  22. Stimulation of Typical Rate Process.............................. 356
    a. Basic Equation.. 356
    b. Number of Transitions per Atom.............................. 358
    c. Disordering.. 360
    d. Values of ρ and ρ'.................................. 360
    e. Sources of Discrepancy....................................... 361
  23. Re-Establishment of Order.. 363
  24. Generation of Frenkel Defects.................................... 364
  25. The Liberation Process... 366
  26. Stresses Around the Displacement Spike; Plastic Flow.............. 368
    a. The Stress Field.. 368
    b. Generation of Dislocation Rings.............................. 369
    c. Persistence of Rings.. 370
    d. Possibility of Disordering in Alloys.......................... 372
    e. Summary.. 373
  27. Establishment of Equilibrium between Conduction Electrons and
     Lattice... 375
    a. Nearly Free Electrons.. 375
    b. d- or f-Shell Electrons.................................. 377
VII. The Simple Theory of Displacement................................. 378
  28. Number of Primary Atoms Displaced............................ 378
    a. Relativistic Case; Fast Electrons............................. 379
  29. Number of Secondary Atoms Dislodged per Primary.............. 380
    a. Hard-Sphere Collisions....................................... 380
    b. Inclusion of Interchange Collisions........................... 383
    c. Continuously Variable Probability of Ejection................. 385
    d. Rutherford Range.. 388
    e. Range of Electron Excitation................................. 389
  30. Estimates of the Number of Displacements and Disordered Atoms.... 390
VIII. Behavior of Metals.. 391
  31. Introduction.. 391
  32. Theoretical Treatment of Scattering of Conduction Electrons........ 393
  33. Huntington's Calculations; Quenching of Defects in Gold............ 395
    a. Huntington's Calculations.................................... 395
    b. The Quenching Experiments of Kauffman and Koehler........... 397

SEI

SEITZ, Frederick

34. An Analysis of Observed Radiation Effects in Copper............... 398
    a. Threshold Energy... 398
    b. Deuteron Bombardment..................................... 399
    c. Electron Bombardment..................................... 402
    d. Expansion Measurements................................... 403
    e. Cooper's Analysis of Radiation Annealing.................... 405
    f. Relative Damage Observed in Copper, Silver, and Gold.......... 407
    g. Pile Irradiation of Copper and Gold........................ 408
35. Annealing Experiments.. 411
    a. Annealing Peaks near 30°K................................. 411
    b. Annealing above 90°K..................................... 415
    c. Annealing of Electron-Bombarded Copper.................... 420
    d. Recovery of Expansion of Copper........................... 421
    e. Nature of Activation Energy near 0.69 eV.................... 422
36. Experiments with Other Pure Monatomic Metals.................... 424
    a. Comparison of Iron, Cobalt, and Nickel...................... 424
    b. Bombardment of Tungsten with High-Energy Protons.......... 425
37. Behavior of Ordered Alloys....................................... 425
    a. Bombardment of Cu_3Au and CuZn with Nucleons............. 425
    b. Bombardment with Electrons............................... 427
    c. Neutron Bombardment of Ni_3Mn........................... 430
    d. Importance of Displacement Spikes.......................... 430
    e. Acceleration of Ordering.................................. 431
38. Evidence for Interstitial Migration............................... 432
IX. Behavior of Valence Crystals....................................... 433
39. Experiments with Germanium..................................... 433
    a. Displacement Threshold................................... 435
    b. Displacements per Electron at Liquid Nitrogen Temperature....... 436
    c. Electron Bombardment at Room Temperature................. 438
    d. Alpha-Particle Bombardment at Room Temperature.............. 440
    e. Deuteron Bombardment at −70°C........................... 440
    f. Annealing above Room Temperature......................... 441
40. Kohn's Calculations... 441
X. Ionic Crystals... 442
41. Expansion of Lithium Fluoride.................................... 442

SELLMYER, D.J.   VOLUME 33, 1978

Electronic Structure of Metallic Compounds and Alloys: Experimental Aspects

D.J. SELLMYER

Behlen Laboratory of Physics, University of Nebraska, Lincoln, Nebraska

I.	Introduction	83
II.	Theoretical Considerations	86
	1. Electron States in Ordered Alloys	86
	2. Simple Models for Disordered Alloys	92
	3. A Survey of Recent Theories and Calculations for Disordered Alloys	102
III.	Survey of Experimental Methods	107
	4. Equilibrium Properties	107
	5. Optical Properties	112
	6. Soft X-Ray Spectroscopy	118
	7. Photoemission Spectroscopy	132
	8. Fermi Surface Images in Inelastic and Diffuse Scattering	142
	9. Positron Annihilation	148
	10. High-Field Fermi Surface Methods	152
IV.	Electron States in Ordered Alloys and Compounds	165
	11. sp-Metallic Compounds	166
	12. Transition Metal Compounds	177
	13. Rare Earth, Chalcogenide, Oxide, and Other Compounds	189
V.	Electron States in Disordered Alloys	205
	14. Fermi Surface Measurements in Primary Solid Solutions	205
	15. sp-Metallic Alloys	213
	16. Noble Metal Alloys	215
	17. Transition Metal Alloys	226
	18. Intermediate Phases, Hydrides, and Amorphous Alloys	235
VI.	Conclusions	247

SHA

SHAM, L.J. VOLUME 15, 1963

The Electron–Phonon Interaction

L.J. SHAM AND J.M. ZIMAN

Cavendish Laboratory, University of Cambridge, Cambridge, England

"Don't let us make imaginary evils, when
you know we have so many real ones to
encounter."

—Oliver Goldsmith.

I. Formal Theory... 223
    1. Introduction via Diffraction Theory................................ 223
    2. The Debye–Waller Factor, and Multiphonon Processes................ 229
    3. The Adiabatic Approximation...................................... 233
    4. The Bloch Formula.. 236
    5. The Rigid-Ion Approximation...................................... 239
    6. Deformation Transformations...................................... 241
    7. The Deformation Potential.. 247
II. Calculations with Rigid Ions... 251
    8. The Orthogonalized-Plane-Wave Method............................. 251
    9. The Pseudopotential.. 254
    10. The Cancellation Theorem.. 257
    11. N Processes and U Processes................................. 262
III. Screening Effects... 268
    12. The Hartree Approximation....................................... 268
    13. Screening and Charge Shift...................................... 271
    14. The Bardeen Formula... 275
    15. Quasi-Particle Formulation for Correlation and Exchange......... 278
    16. The Deformation Potential in a Metal............................ 284
IV. Observation of the Electron-Phonon Interaction....................... 286
    17. Ultrasonic Attenuation.. 286
    18. Electrical Resistivity.. 288
    19. A Digression on Liquid Metals................................... 290
    20. Superconductivity... 295

SHAW, T.M.: *see* BEYERS, R.

Applications of Neutron Diffraction
to Solid State Problems

C.G. SHULL AND E.O. WOLLAN

Oak Ridge National Laboratory, Oak Ridge, Tennessee

  I. Introduction.. 138
 II. General Considerations... 138
III. Crystal Studies.. 149
     1. Hydrogen-Containing Structures..................................... 149
        a. Hydrides.. 150
        b. Ice... 151
        c. Potassium Bifluoride.. 153
        d. Ammonium Halides.. 154
        e. Other Hydrogen-Containing Structures............................ 156
     2. Ferroelectric Structures... 158
        a. Titanates... 159
        b. KH_2PO_4.. 160
     3. Other Structural Studies... 162
     4. Liquid and Amorphous Structures.................................... 163
     5. Inelastic Scattering Effects....................................... 167
 IV. Magnetic Studies.. 177
     6. Transition Group Elements and Alloys............................... 177
        a. The $3d$ Group Metals... 177
        b. Elements of the $4d$ and $5d$ Series............................ 181
        c. Binary Alloys of the $3d$ Group................................. 182
        d. Rare-Earth Metals... 183
     7. Antiferromagnetic Structures....................................... 184
        a. Iron Group Compounds with NaCl Structure........................ 184
        b. Some Compounds with Hexagonal Structure......................... 190
        c. Fluoride Structures... 190
        d. α-Fe_2O_3 and α-Cr_2O_3....................... 190
        e. Manganese... 191
        f. Perovskites... 191
     8. Ferrimagnetic Compounds.. 191
     9. Perovskite-Type Compounds.. 195
    10. Paramagnetic Scattering and Order-Disorder Transition Phenomena.... 199
    11. Magnetic Form Factors.. 208
    12. Indirect Exchange Mechanisms....................................... 214

SIN

SINGH, Jai VOLUME 38, 1984

The Dynamics of Excitons

JAI SINGH

Research School of Chemistry, Australian National University, Canberra, Australia

I.   Introduction 295
    1. Excitons 296
    2. Theoretical Methods for Crystalline Solids 297
II.   Theory of Excitons 301
    3. The Hamiltonian in Second Quantization 301
    4. Wannier Excitons 304
    5. The Wannier Exciton Hamiltonian 314
    6. Frenkel Excitons 315
    7. The Frenkel Exciton Hamiltonian 321
    8. Motion of Frenkel Excitons 321
    9. Critical Observations on the Frenkel Exciton Theory 324
    10. Comparison of the Wannier and Frenkel Excitons 326
III.   Exciton–Phonon Interactions 328
    11. The Wannier Exciton–Phonon Interaction Operator 329
    12. The Frenkel Exciton–Phonon Interaction Operator 332
IV.   Composite Exciton–Phonon States 336
    13. The Energy Eigenvalue and Eigenvector of the Composite
        Exciton–Phonon States (0–0 Phonon Transition) 337
V.   Exciton Reactions 341
    14. The Reaction Rate 342
    15. Binding and Decay of Excitons in Pure Crystals 342
    16. Exciton–Exciton Collisions 352
    17. Fission and Fusion of Excitons 358
    18. Exciton–Charge Carrier Interactions 363
    19. Rates of the Nonradiative Decay of Excitons 368

SINGWI, K.S.:   *see* KOTHARI, L.S.

SINGWI, K.S.   VOLUME 36, 1981

Correlations in Electron Liquids

K.S. SINGWI

Department of Physics and Astronomy, Northwestern University, Evanston, Illinois

AND

M.P. TOSI

International Center for Theoretical Physics, Trieste, Italy

I.	Introduction	177
II.	Homogeneous Electron Liquid	179
	1. Density Response and Dielectric Function	179
	2. Limiting Behaviors and Sum Rules	186
	3. Equations of Motion for Density Matrices	191
	4. Mean-Field Theory	203
III.	Approximate Schemes	207
	5. Self-Consistent Theory and Local Field Correction	207
	6. Static Dielectric Function	220
	7. Dynamical Behavior of the Electron Liquid	235
IV.	Some Applications	244
	8. Two-Component Quantum Plasma	244
	9. Spin Correlations	254
	10. Two-Dimensional Electron Fluid	257
	11. Dielectric Response of Semiconductors and Insulators	260
	Note Added in Proof	266

SLACK, Glen A.   VOLUME 34, 1979

The Thermal Conductivity of Nonmetallic Crystals

GLEN A. SLACK

General Electric Research and Development Center, Schenectady, New York

I.	Introduction	1
II.	Absolute Value of the Thermal Conductivity	2
	1. Previous Calculations, Acoustic Phonons	2
	2. Rare-Gas Crystals ($n = 1$)	5
	3. Adamantine Crystals ($n = 2$)	9
	4. Rocksalt Structure Crystals ($n = 2$)	17
	5. Heat Transport by Optic Phonons	21
	6. Other Crystals ($n \geq 3$)	23
	7. Generalizations	31
	8. Longitudinal and Transverse Phonons	33

SLA–SMI

SLACK, Glen A.

III. Volume Dependence of the Thermal Conductivity 34
    9. Hydrostatic Stress 34
    10. Uniaxial Stress 36
IV. Temperature Dependence of the Thermal Conductivity 37
    11. Thermal Expansion Effects 37
    12. Scattering by Optic Phonons 39
V. Thermal Conductivity at the Melting Point 44
    13. Theory for Liquids 44
    14. Volume Dependence for Liquids 47
    15. Volume Dependence for Disordered Solids 49
    16. Grüneisen Parameter for Liquids 51
    17. Solid–Liquid Transition 52
    18. Summary 57
VI. Minimum Thermal Conductivity 57
    19. Rare-Gas Crystals ($n = 1$) 57
    20. Adamantine and Rocksalt Structure Crystals ($n = 2$) 59
    21. Fluorite Structure Crystals ($n = 3$) 63
    22. Other Crystals ($n \geq 4$) 66
    23. Liquids 69
VII. Further Problems 70

SMITH, Charles S. VOLUME 6, 1958

Macroscopic Symmetry and Properties of Crystals

CHARLES S. SMITH

Case Institute of Technology, Cleveland, Ohio

I. Introduction 175
II. Physical Quantities and Properties 178
III. Crystal Symmetry and Classification 181
    1. Symmetry Operators 182
    2. Point Groups 183
    3. Crystal Classification 186
    4. Physical Properties and Symmetry 191
IV. First Rank Tensor Properties 196
    5. Spontaneous Polarization 197
    6. The Pyroelectric Effect 200
V. Second Rank Tensor Properties 207
    7. Transport Properties—Electrical Resistivity 208
    8. Thermodynamic Properties—Dielectric Susceptibility 217
    9. Tensor-Scalar Properties—Thermal Expansion 220
VI. Third Rank Tensor Properties 223
    10. Piezoelectric Effect 224
VII. Fourth Rank Tensor Properties 231
    11. Piezoresistivity 232
    12. Elasticity 241

SOOD, Ajay K.   VOLUME 45, 1991

Structural Ordering
in Colloidal Suspensions

AJAY K. SOOD

Department of Physics
Indian Institute of Science
Bangalore, India

I.	Introduction	2
	1. Colloids in General and the Aims of This Review	2
	2. Model Colloidal Systems.	3
	3. Comparison with Conventional Atomic Systems	6
II.	Synthesis and Characterization of Colloids.	8
	4. Emulsion Polymerization	8
	5. Determination of Surface Charge and Surface Potential	10
III.	Experimental Scene	12
	6. Phase Diagram	12
	7. Growth Kinetics and Time Evolution.	18
IV.	Interparticle Interaction Potentials.	19
	8. DLVO Theory.	19
	9. Electrostatic Interaction with Both Repulsion and Attraction.	27
V.	Theoretical Aspects of Equilibrium Phases	30
	10. Fluid–Crystal and bcc–fcc Phase Transitions	30
VI.	Liquidlike Order, Polydispersity, Binary Mixtures, and Glass	41
	11. The Liquidlike Short-Ranged Order	41
	12. Polydispersity Effects	52
	13. Colloidal Alloys in Binary Mixtures	55
	14. Colloidal Glasses	56
VII.	Colloidal Crystals in Two Dimensions	62
	15. Structure of 2D Crystals	63
	16. Melting of 2D Colloidal Crystals.	64
VIII.	Elastic Properties and Shear Flow	66
	17. Elastic Properties	66
	18. Shear Flows	68
IX.	Miscellaneous and Conclusions.	70
	19. Some Topics Not Covered	70
	20. A Few Unsolved Problems	71
	21. Summary	72
	Acknowledgments	73
	Note Added in Proof	73

SOR

SORBELLO, Richard S.   VOLUME 51, 1997

Theory of Electromigration

RICHARD S. SORBELLO

Department of Physics, University of Wisconsin–Milwaukee, Milwaukee, Wisconsin

I.	Introduction	159
II.	Driving Force for Electromigration	163
	1. Phenomenological Equations of Irreversible Thermodynamics	163
	2. Effective Valence	170
	3. Microscopic Forces and the Adiabatic Picture for Electromigration	174
	4. Theoretical Framework for Microscopic Forces in Electromigration	177
III.	Theoretical Calculations	181
	5. Ballistic Model	181
	6. Bosvieux–Friedel Analysis	185
	7. Pseudopotential-Based Analysis of the Wind Force	191
	8. Landauer's Analysis	200
	9. Das–Peierls Analysis	205
	10. Density Matrix Analysis	208
	11. Linear Response Theory	212
	12. Multiple Scattering Theory Calculations of the Wind Force	219
	13. Time-Dependent Density-Functional Analysis	223
IV.	Beyond the Adiabatic Picture	224
	14. Dynamical Effects in Electromigration	224
	15. Quantum Diffusion of Hydrogen in Metals	226
V.	Conclusion	228
	16. Summary and Outlook	228

SPAEPEN, Frans   VOLUME 47, 1994

Homogeneous Nucleation and the Temperature Dependence of the Crystal-Melt Interfacial Tension

FRANS SPAEPEN

Division of Applied Sciences, Harvard University, Cambridge, Massachusetts

I.	Introduction	1
II.	The Equilibrium Interface	5
	1. Temperature Dependence of the Interfacial Tension: Excess Interfacial Entropy	5
III.	The Interface in an Undercooled Melt	8
	2. The Infinitely Thin Interface	8
	3. Structure and Properties of the Physical Interface	10
	4. Interfacial Tension of the Physical Interface	12
	5. A Simple Model: The Uniform Interface	14
	6. The Simple Model at Low Undercooling	17
IV.	Application to the Analysis of Homogeneous Nucleation Experiments	20
	7. Reanalysis of the Data for Mercury	20
	8. Interpretation of the Temperature Dependence of the Interfacial Tension	24
V.	Conclusions	27
	Acknowledgments	28
	Appendix A: Calculation of the Interfacial Tension as an Excess Quantity for a Curved Interface	28
	Appendix B: Calculation of the Interfacial Tension for a General Spatial Variation of the Free Energy	30

SPAEPEN, Frans: *see* NELSON, David R.

SPE–STE

SPECTOR, Harold N.   VOLUME 19, 1966

Interaction of Acoustic Waves and Conduction Electrons

HAROLD N. SPECTOR

IIT Research Institute, Chicago, Illinois

I. Introduction... 291
II. Theory of the Interaction.. 293
III. Calculation of the Conductivity Tensor................................. 306
   1. Semiclassical Treatment.. 307
   2. Quantum Treatment... 317
IV. Calculation of the Absorption Coefficient.............................. 320
   3. Power Transferred.. 320
   4. Dispersion Relations... 339
   5. Acoustoelectric Effect... 343
V. Electron-Phonon Interaction as a Tool.................................. 349
   List of Symbols.. 356

STERN, Frank   VOLUME 15, 1963

Elementary Theory of the Optical Properties of Solids

FRANK STERN

United States Naval Ordnance Laboratory, White Oak, Silver Spring, Maryland

I. Electromagnetic Preliminaries... 300
   1. Introduction... 300
   2. Maxwell's Equations... 301
   3. Conservation Equations.. 304
   4. Plane Waves... 305
   5. Inhomogeneous Case... 307
   6. Energy Density.. 308
   7. Sources, Potentials, and the Wavelength-Dependent Dielectric Constant.. 310
   8. Alternative Treatments of the Magnetization........................ 312
II. Reflection and Transmission... 314
   9. Boundary Conditions... 314
   10. Amplitude Reflection Coefficients.................................. 316
   11. Snell's Law and Generalizations................................... 317
   12. Energy Flow.. 318
   13. Reflectivity... 320
   14. Total Reflection... 322
   15. Transmission of Isotropic Radiation............................... 322
   16. Thin Films... 324

STERN, Frank

III. Dispersion Relations... 327
   17. Dispersion Relations for the Dielectric Constant...................... 328
   18. Dispersion Relations for the Index of Refraction...................... 331
   19. Qualitative Consequences.. 332
   20. Phase Shift Dispersion Relation.................................... 333
   21. Numerical Procedures for Calculating the Phase...................... 337
   22. Extrapolation Procedures.. 338
   23. Derivation of Sum Rules... 340
IV. Optical Properties of Simple Systems................................. 342
   24. The Effective Field.. 342
  A. The Free-Electron Gas... 344
   25. General Considerations.. 344
   26. Conductivity and Mobility... 346
   27. Optical Properties in Various Frequency Ranges...................... 347
   28. Reflectivity Minimum.. 350
   29. Energy Density... 350
  B. Optical Modes in Ionic Crystals.................................... 351
   30. Phenomenological Discussion....................................... 351
   31. Atomic Theory of Optical Modes.................................... 354
   32. Longitudinal and Transverse Modes; Effective Charge................ 356
   33. Reflectivity.. 359
   34. Free-Carrier Effects.. 361
V. Quantum Theory of Absorption and Emission............................ 364
   35. Absorption... 364
   36. Blackbody Radiation and Spontaneous Emission...................... 367
   37. Optical Absorption for a Simple Model............................. 370
   38. Free-Carrier Absorption... 374
   39. Oscillator Strengths.. 375
VI. Wavelength-Dependent Dielectric Constant of a Free-Electron Gas........ 379
   40. Introduction... 379
   41. Calculation of the Conductivity................................... 380
   42. Verification of the Sum Rule...................................... 383
   43. Calculation of the Dielectric Constant............................. 385
   44. Dielectric Constant with no External Currents...................... 386
   45. Surface Impedance and Anomalous Skin Effect...................... 389
   46. Real Crystals.. 393
VII. Interactions of Charged Particles.................................... 394
   47. Longitudinal Dielectric Constant of a Free-Electron Gas............. 394
   48. Screening of a Static Point Charge................................. 398
   49. Energy Loss of a Moving Charged Particle.......................... 404
   50. Forces between Charged Particles.................................. 406
   Acknowledgments.. 408

STÖCKMANN, F.:   *see* HEILAND, G.

STR–STU

STRONG, H.M.: *see* BUNDY, F.P.

STROUD, D.: *see* ASHCROFT, N.W.

STROUD, David: *see* BERGMAN, David J.

STURGE, M.D. VOLUME 20, 1967

The Jahn–Teller Effect in Solids

M.D. STURGE

Bell Telephone Laboratories, Incorporated, Murray Hill, New Jersey

I. Introduction 92
   1. Crystal Defects and Transition Metal Ions in Insulating Crystals 93
   2. Collective Coordinates and the "Quasi-Molecular" Model 100
   3. The Jahn–Teller Theorem 106
   4. The Static and Dynamic Jahn–Teller Effect: Some Orders of Magnitude . . 113
II. The Jahn-Teller Effect in Doubly Degenerate Electronic States 115
   5. The Static Problem in an Octahedral Complex 115
   6. The Dynamic Problem: Vibronic Energy Levels 122
   7. The Transition from Static to Dynamic Jahn–Teller Effect: Motional
      Narrowing 126
   8. Tunneling and Spin Resonance at Low Temperatures 129
   9. Acoustic Consequences of the Jahn–Teller Effect 134
  10. *E* Terms in Tetrahedrally Coordinated Ions 140
  11. Centers with Less Than Cubic Symmetry 141
  12. Square Planar Complexes 143
  13. Cooperative Jahn–Teller Effects 146

STURGE, M.D.

III. The Jahn–Teller Effect in Triply Degenerate States...................... 151
 14. An Octahedral Complex in a ^1T State, Coupling Only to ϵ_g Modes....... 152
 15. Angular Momentum and Weak Spin-Orbit Coupling: The Ham Effect.... 154
 16. Higher-Order Terms in the Ham Effect............................. 159
 17. Jahn–Teller Coupling to τ_{2g} Distortions in Octahedral Complexes....... 162
 18. Tetrahedral Complexes.. 167
 19. Vacancies and Other Defects in the Diamond Structure............. 170
IV. Optical Transitions Involving Jahn–Teller Distorted States........ 178
 20. The Configuration-Coordinate Model............................ 178
 21. Excitation of Nontotally Symmetric Modes of Vibration........... 182
 22. Transitions from Doubly Degenerate States...................... 186
 23. Broadband Transitions to Doubly Degenerate States.............. 188
 24. No-Phonon Transitions Involving Doubly Degenerate States....... 192
 25. Optical Transitions Involving Triply Degenerate States........... 193
 26. The Case of Strong Spin-Orbit Coupling: F-Bands in Cesium Halides.... 194
 27. Band Splittings Due to τ_2 Vibrations: KCl: Tl$^+$ and Isoelectronic Centers.. 198
 28. Fine Structure and the Ham Effect in Optical Spectra of Transition Metal Ions.. 201
 29. Conclusions... 210

SWENSON, C.A.   VOLUME 11, 1960

Physics at High Pressure

C.A. SWENSON

Institute for Atomic Research and Department of Physics, Iowa State University, Ames, Iowa

I. Introduction... 41
II. Experimental Methods.. 44
 1. The Pressure Scale.. 44
 2. Techniques—Mainly Room Temperature........................... 47
 3. Low-Temperature Techniques..................................... 60
III. Results.. 73
 4. PVT Data for Solids... 73
 5. Electrical and Magnetic Measurements........................... 103
 6. Optical Measurements... 130
 7. Diffusion.. 135
 8. Miscellaneous.. 137
 9. Fluids: PVT and Transport Data................................ 142

TAKAGI, Yutaka: *see* MUTO, Toshinosuke

TARDIEU, Annette: *see* CHARVOLIN, Jean

THOMAS, G.A.: *see* HENSEL, J.C.

THOMSON, Robb   VOLUME 39, 1986

Physics of Fracture

ROBB THOMSON

National Bureau of Standards
Washington, D.C.

I.	Introduction	2
II.	Observational Background and Initial Concepts	7
	1. Macroscopic Fractographic Observations	7
	2. Microscopic Observations	9
	3. Prototype Fractures	10
	4. Fracture Mode and Crack Dislocation	18
	5. The Observational Basis for Fracture Science	18
III.	The Elastic Fields of Cracks and Dislocations	21
	6. Introduction	21
	7. General Equations and Boundary Conditions	22
	8. Antiplane Strain. The Mode-III Crack	25
	9. Plane Strain and Plane Stress	28
	10. Dislocation Stress Fields	31
	11. Wedge Cracks	36
	12. Anisotropic Elasticity	38
IV.	The Elastic Forces on Cracks and Dislocations	39
	13. Introduction	39
	14. The Force on an Elastic Singularity: Eshelby's Theorem	41
	15. Antiplane Strain. Mode-III and Screw Dislocations	44
	16. Plane Strain: Edge Dislocations and Mode-I and Mode-II Cracks	47
	17. The Wedge Case	52
	18. The J Integral	53
	19. The Griffith Relations	56

THOMSON, Robb

V.	Atomic Structure of Cracks	58
	20. Introduction	58
	21. Lattice Statics	58
	22. Discrete-Enclave Method	64
	23. Molecular Dynamics	65
	24. Force Laws for Defect Calculations	66
	25. Chemical Kinetics	67
	26. Slow Crack Growth Observations	70
VI.	Dislocation Emission from Cracks	71
	27. Introduction	71
	28. Elastic Estimates for Sharp Cracks	72
	29. Thermally Activated Emission	78
	30. Atomic Calculations	80
	31. Experimental Investigations	83
VII.	Dislocation Shielding and Fracture Toughness	87
	32. Introduction	87
	33. Equilibrium Conditions for the Three Prototype Crack Classes	89
	34. One-Dimensional Dislocation Distribution	95
	35. The Fracture Criterion in Cleavage	99
	36. Emission-Controlled Shielding	101
	37. Shielding by Externally Produced Dislocations	105
	38. Mode-I and Mixed-Mode Problems	107
	39. Growing Cracks	113
	40. Time-Dependent Cleavage/Emission Considerations	119
	Appendix	120

THOMSON, R.: *see* CARLSSON, A.E.

TIN–TOS

TINKHAM, M.
VOLUME 42, 1989

Physical Properties
of the New Superconductors

M. TINKHAM AND C.J. LOBB

Physics Department and Division of Applied Sciences
Harvard University
Cambridge, Massachusetts

1. Introduction 91
I. Magnetic and Resistive Properties 92
  2. The Importance of Granularity 92
  3. Effective Medium Parameters 96
  4. Relationship Between Granular and Continuum Models 103
  5. Flux Creep and Irreversibility 107
II. Other Physical Properties 118
  6. Isotope Effect 118
  7. Energy Gap 122
  8. Electronic Specific Heat 131
  9. Concluding Remarks 133
  Notes Added in Proof 134

TOSI, Mario P.
VOLUME 16, 1964

Cohesion of Ionic Solids in the Born Model

MARIO P. TOSI

Argonne National Laboratory, Argonne, Illinois

I. Introduction 1
II. Cohesive Energy 6
  1. Cohesive Energy and Equation of State of Cubic Solids 6
  2. Crystal Structure of Simple Ionic Solids 9
  3. Electrostatic Lattice Energy and Madelung Constant 13
  4. Van der Waals Energy 29
  5. Born Repulsive Energy 35
  6. Cohesive Energy of the Alkali Halide Crystals 42
  7. Cohesive Energy of Some "Essentially Ionic" Crystals 59
  8. Relative Stability of Different Crystal Structures for Ionic Compounds .. 68

TOSI, Mario P.

III. Ionic Radii... 74
    9. Foundations of the Notion of Crystal Radius........................ 74
    10. Standard Crystal Radii... 77
    11. Determination of Crystal Radii by the Born Model.................. 84
    12. Crystal Radii and Apparent Sizes of Ions in Crystals.............. 88
IV. Surface Energy... 92
    13. Surface Energy: Definition and Measurement....................... 92
    14. Specific Face Energy.. 96
    15. Finiteness Effects in the Surface Energy......................... 105
    Appendix A. Evaluation of Electrostatic Lattice Potentials by the Ewald
        Method.. 107
    Appendix B. A Review of Methods of Lattice Summation................. 113

TOSI, M.P.: *see* SINGWI, K.S.

TURNBULL, David   VOLUME 3, 1956

Phase Changes

DAVID TURNBULL

Research Laboratory, General Electric Co., Schenectady, New York

 I. Introduction.. 226
 II. Thermodynamics of Phase Changes...................................... 226
    1. Principles of Phase Equilibria...................................... 226
      a. General.. 226
      b. One-Component Systems... 228
      c. Two-Component Systems... 233
    2. Thermodynamics of Interfaces....................................... 236
      a. Interfaces between Fluid Phases.................................. 236
      b. Interfaces between Solid and Fluid Phases........................ 238
III. Remarks on the Theory of Phase Stability............................. 240
    3. General Considerations.. 240
    4. Melting.. 241
    5. Transitions between Crystalline Phases (Polymorphism).............. 245
    6. Stability of Solutions... 248
IV. Theory for Phase Change Kinetics...................................... 252
    7. Formal Theory.. 252
    8. General Theory of Nucleation....................................... 256
      a. Nucleation of Liquid Droplets in a Vapor......................... 257
      b. Experience on the Kinetics of Vapor-Liquid Transitions........... 263

TURNBULL, David

9. Theory of Nucleation in Condensed Systems............................ 266
   a. Homogeneous Nucleation at Constant Composition................ 266
   b. Transient Nucleation... 268
   c. Homogeneous Nucleation of Precipitate......................... 269
   d. Interfaces between Crystalline Phases and Transformation Strains... 272
   e. Heterogeneous Nucleation...................................... 275
10. Theory of Growth Processes.. 279
V. Interpretation of Experience on Phase Change Kinetics................. 281
11. Liquid → Crystal Transitions at Constant Composition.............. 281
   a. General.. 281
   b. Homogeneous Nucleation....................................... 283
   c. Heterogeneous Nucleation..................................... 286
   d. Growth.. 289
   e. Implications to Theory of Liquid State and Melting............... 291
12. Transitions, at Constant Composition, between Crystalline Phases..... 293
   a. General.. 293
   b. Martensite Transformations.................................... 295
13. Precipitation of Crystals.. 298
   a. From Liquid Solutions... 298
   b. General Precipitation from Solid Solutions...................... 299
   c. Cellular Precipitation from Solid Solutions..................... 301
   d. Guinier-Preston Zone Formation................................ 303
VI. Conclusions... 306

TYLER, W.W.: *see* NEWMAN, R.

VAMANU, D.: *see* CORCIOVEI, A.

van HOUTEN, H.: *see* BEENAKKER, C.W.J.

van LINT, V.A.J.: *see* NICHOLS, D.K.

VINK, H.J.: *see* KRÖGER, F.A.

WALLACE, Duane C.   VOLUME 25, 1970

Thermoelastic Theory of Stressed Crystals and Higher-Order Elastic Constants

DUANE C. WALLACE

Sandia Laboratories, Albuquerque, New Mexico

I. Introduction	302
II. Thermoelasticity of Stressed Materials	305
1. Thermodynamic Functions	305
2. Derivation of the Basic Equations	309
3. Small Amplitude Elastic Wave Propagation	314
4. Harmonic Generation	318
5. Attenuation of Elastic Waves	324
6. Stress–Strain Relations and Compressibility	328
7. Case of Initial Isotropic Pressure	334
III. Expansions for Small Initial Stress	336
8. Expansion of Elastic Constants	336
9. Expansion of Strains and Propagation Matrices	339
10. Stress Derivatives of the Transit Time	343
IV. Crystal Symmetry Properties	347
11. Schemes of Independent Elastic Constants	347
12. Cubic I Crystal under Isotropic Pressure	351
13. Tetragonal I Crystal under Uniaxial Stress	357
14. Isotropic Material	360
V. Measurement of Third-Order Elastic Constants	363
15. Wave Propagation in Stressed Crystals—Tables of Results	363
16. Wave Propagation in Stressed Crystals—Interpretation of Experiments.	368
17. Harmonic Generation—Tables of Results	372
18. Harmonic Generation—Interpretation of Experiments	376

WAL

WALLACE, Duane C.

VI. Fourth-Order Elastic Constants of Cubic Crystals...................... 380
   19. Application of Isotropic Pressure or Uniaxial Stress................. 380
   20. Application of Isotropic Pressure plus Uniaxial Stress............... 385
   21. Harmonic Generation in Stressed Crystals........................ 389
VII. Interpretation of Measurements in Terms of Atomic Theories............ 391
   22. Application of Lattice Dynamics................................. 391

WALLACE, Philip R.   VOLUME 10, 1960

Positron Annihilation in Solids and Liquids

PHILIP R. WALLACE

McGill University, Montreal, Canada

Introduction.. 1
I. The Annihilation Process.. 3
II. Measurements of Lifetime for Positron Annihilation and Their Interpretation.. 8
III. Interpretation of the Lifetime Measurements.......................... 16
IV. The Short Lifetime: Positron Decay in Type A Materials............... 22
   1. Angular Correlation of the Emitted γ Rays......................... 22
   2. Theory of Annihilation in Metals............................... 36
   3. Annihilation in Ionic Crystals.................................. 41
   4. Annihilation in Oriented Graphite.............................. 43
   5. Annihilation in Superconductors............................... 44
   6. Mobility of Positrons in Diamond.............................. 45
   7. Three-Quantum Annihilation Rates............................. 46
V. The Long Lifetime and Variations in Positronium Formation........... 46
   8. Annihilation in Gases.. 46
   9. Annihilation in Liquid Helium.................................. 53
   10. Variations in Long Lifetime and Its Intensity—Effect of Temperature
   and Density... 57
   11. Quenching by Magnetic Fields................................. 62
   12. The Effect of Impurities on Positron Annihilation.............. 64

WALSH, J.M.:   *see* RICE, M.H.

WANG, Xiaoyi: *see* OSGOOD, Richard M., Jr.

WEAIRE, D.:   *see* HEINE, Volker

WEAIRE, D.:   *see* Miller, A.

WEAIRE, D.:   *see* WOOTEN, F.

WEAIRE, D.   VOLUME 50, 1996

Some Fundamentals
of Grain Growth

D. WEAIRE AND S. MCMURRY

Department of Physics, Trinity College, Dublin, Ireland

SOME FUNDAMENTALS OF GRAIN GROWTH   1
   I. Introduction   1
  II. Ideal Grain Growth   3
      1. Two Dimensions.   3
      2. Three Dimensions.   7
 III. Some Elementary Solutions.   7
 IV. The Motion of Boundary Intersections in 2D and 3D   10
      3. Two Dimensions.   10
      4. Three Dimensions.   12
  V. The Law of Mullins and von Neumann   13
      5. Two Dimensions.   13
      6. Three Dimensions.   17

WEA

WEAIRE, D.

VI. Asymptotic Properties and the Fate of Small Cells 18
   7. Scaling 18
   8. Small Cells 19
   9. Three Dimensions 19
VII. Approximate Models 19
VIII. Three-Dimensional Simulation 24
IX. Analogous Systems 24
X. Separated Grains: Lifshitz-Slyozov Theory 26
   10. Ostwald Ripening 26
   11. The Mathematical Model 28
   12. The Asymptotic Analysis 30
XI. Conclusion 35

WEAVER, J.H.   VOLUME 48, 1994

Solid State Properties of Fullerenes and Fullerene-Based Materials

J.H. WEAVER

D.M. POIRIER

Department of Materials Science and Chemical Engineering
University of Minnesota, Minneapolis, Minnesota

  I. Introduction 1
    1. Chapter Overview and Goals 1
    2. Historical Perspectives 4
    3. The Fullerene Family 8
    4. Fullerene Formation 10
  II. Fullerenes in Gas Phase and Solution 13
    5. Fullerenes in Gas Phase 14
    6. Fullerenes in Solution 16
  III. Fullerenes in the Solid State 18
    7. Solid State Structures and Ordering Transitions 18
    8. Thin Films, Thin Film Formation, and Substrate Interactions 27
    9. Electronic Structure 35
    10. Vibrons and Phonons 53
    11. Fullerene Oxidation, Modification, and Polymerization . . . 56
  IV. Doped Fullerenes: Fullerides 60
    12. Fulleride Crystal Structure 63
    13. The $K\text{-}C_{60}$ Phase Diagram 67
    14. Electrical Resistivity and Granular Metal Behavior 69
    15. Fulleride Electronic Structures 71
    16. Fulleride Superconductivity 82
    17. Alkaline-Earth Fullerides 86
  V. Endofullerenes 89
  VI. Tubes, Capsules, and Onions 95
  VII. Summary and Prospects 106
    Acknowledgments 107

WEBB, M.B.   VOLUME 28, 1973

Elastic Scattering of Low-Energy Electrons from Surfaces*

M.B. WEBB AND M.G. LAGALLY

University of Wisconsin, Madison, Wisconsin

I. Introduction 302
II. Atomic Scattering 306
III. Kinematic Description of Diffraction............................. 313
    1. Penetrating Radiation... 313
    2. Diffraction from a Crystal.................................... 315
    3. Attenuated Radiation.. 317
    4. Inner Potential 320
    5. Thermal Motion... 321
IV. Experimental Aspects.. 323
    6. Apparatus 323
    7. Instrument Response... 326
    8. Lorentz Factor 329
    9. Notation and Experimental Capabilities........................ 331
V. Some General Conclusions from Experimental Results................. 332
    10. Integrated Intensity 333
    11. Angular Distribution and Spot Patterns 334
    12. Intensity Profiles.. 335
    13. Multiple Scattering: Scattering from Liquid Hg 341
    14. Scattering Factor 347
    15. Attenuation 349
VI. Dynamic Theories... 357
    16. Extensions of the Kinematic Theory 357
    17. Static Potentials 360
    18. Inclusion of Inelastic Scattering........................... 365
    19. Realistic Calculations...................................... 369
VII. Effect of Temperature 380
    20. Zero-Phonon Scattering: Debye–Waller Factor 385
    21. Thermal Diffuse Scattering.................................. 389
    22. Energy Shift: Thermal Expansion............................. 392
VIII. Surface Crystallography 395
    23. Structure Determination from Dynamic Calculations........... 396
    24. Averaging Techniques.. 398
IX. Conclusions 404

WEG

WEGER, M.   VOLUME 28, 1973

Some Lattice and Electronic Properties
of the β-Tungstens

M. WEGER AND I.B. GOLDBERG

The Racah Institute of Physics, The Hebrew University, Jerusalem, Israel

I. Introduction 2
  1. Experimental 7
    a. The Martensitic Transformation.................................. 7
    b. Susceptibility and NMR Measurements........................... 11
    c. Specific Heat.. 19
    d. Transport Properties... 26
    e. Fermi Surface Measurements..................................... 29
  2. Martensitic Transformations and Ferroelasticity 33
  3. Simple Treatment within the Landau Theory 37
    a. Landau Theory of Second-Order Phase Transitions................. 37
    b. The Martensitic Transformation and the Superconducting Phase
       Transition 40
II. Description of the Martensitic Transformation from First Principles 43
  4. Models Describing Lattice Instabilities 43
    a. The Frohlich Model.. 43
    b. The Labbe–Friedel–Barisic Model.............................. 44
    c. The RCA Model.. 49
    d. The Tomonaga Model... 56
    e. The Anderson–Blount and Klein–Birman Models.................. 58
  5. Description of the Martensitic Transformation 62
    a. Expansion of the Energy at 0°K................................ 62
    b. Behavior at Finite Temperatures 66
    c. Application of the Landau Theory............................... 68
    d. Some Difficulties of the Labbe–Friedel–Barisic Model.............. 81
    e. Contributions of States of Various Symmetries to the Elastic Constants   85
    f. Fluctuations 88
    g. Difficulties Associated with the Optical Phonon Instability Model.... 90
III. The Electronic Band Structure of the β-Tungstens 91
  6. The Model of Clogston and Jaccarino............................... 91
  7. The Linear Chain Model 93
  8. The APW Calculations of Mattheiss................................ 99
    a. Description of the APW Method................................ 99

WEGER, M.

WEISS, G.H.: *see* MARADUDIN, A.A.

WEISS, H.: *see* WELKER, H.

WEL

Group III–Group V Compounds

H. WELKER AND H. WEISS

Research Laboratory, Siemens-Schuckertwerke, Erlangen, Germany

I. Introduction.. 1
  1. General Comments... 1
I. Physical Chemical Properties of the III–V Compounds................. 6
  2. Phase Diagrams.. 6
  3. Preparation... 8
  4. Binding Properties.. 11
III. Electrical, Magnetic, and Optical Properties........................ 12
  5. Conductivity and Hall Effect................................ 12
  6. Variations in Resistivity with Magnetic Field............... 33
  7. The Hall Generator.. 39
  8. Thermopotential and Heat Conductivity....................... 42
  9. Infrared Absorption... 49
  10. Lifetime and Recombination on the Surface.................. 58
IV. General Theoretical Considerations................................. 67
  11. Strengthening by Resonance................................. 67
  12. Band Structure and the Ionic Component in Binding.......... 71

The Structures of Crystals

A.F. WELLS

Imperial Chemical Industries Ltd., Manchester, England

I. Some Basic Geometry.. 426
  1. Introduction... 426
  2. Periodic Networks of Connected Points....................... 429
    a. 3-Connected Nets... 429
    b. 4-Connected Nets... 433
    c. 6-Connected Nets... 434
    d. Nets of Higher Coordination Number....................... 434
  3. Space-Filling Arrangements of Polyhedra..................... 435
  4. Open Packings of Polyhedra.................................. 437
  5. The Closest Packing of Spheres.............................. 438
    a. The Closest Packing of Equal Spheres of One Kind......... 438
    b. Close-Packed Arrangements of Atoms of Two Kinds.......... 441

WELLS, A.F.

II. Tetrahedral Coordination of Metal Atoms in Close-Packed Structures.... 442
   6. Survey of Possible Structures...................................... 442
   7. The Nonoccurrence of AB . . . . and $ABAC$ . . . . Structures...... 443
   8. The Fluorite Family of Structures................................. 444
III. Octahedral Coordination of Metal Atoms in Close-Packed Oxides and
   Halides.. 449
   9. Survey of Possible Structures..................................... 449
  10. Structures MX... 451
  11. Structures M_2X_3.. 452
  12. Structures MX_2... 454
    a. Derivation from MX Structures.............................. 454
    b. The Rutile Structure....................................... 456
  13. Structures MX_3... 458
    a. Geometrical Possibilities.................................... 458
    b. The Structures of Trioxides and Trihalides.................... 461
IV. Structures Based on Close-Packed AX_3 Layers....................... 463
  14. Complex Oxides and Halides $A_xB_yX_{3z}$........................... 463
    a. Survey... 463
    b. The Perovskite Structure.................................... 464
    c. The Cryolite Family of Structures............................ 465
  15. Hydroxyhalides $M_2X(OH)_3$.. 467
V. Tetrahedral and Octahedral Coordination in Close-Packed Structures.... 468
  16. The $BiLi_3$, Olivine, and Spinel Structures......................... 468
VI. Structures with Very High Coordination Numbers..................... 470
  17. Types of Coordination Polyhedra................................. 470
  18. The Laves Phases... 472
VII. Structures Based on 3-Connected Nets.............................. 474
  19. Structures of Elements and Simple Compounds..................... 474
    a. Finite Complexes.. 475
    b. Linear Systems... 475
    c. Plane Nets... 475
    d. Three-Dimensional Frameworks............................... 476
  20. Hydrogen-Bonded Structures..................................... 476
VIII. Structures Based on 4-Connected Nets............................. 478
  21. The Diamond and Related Structures............................. 478
  22. Structures Based on More Complex Nets.......................... 480
  23. Hydrogen-Bonded Structures..................................... 482
IX. Structures Based on 6-Connected Nets.............................. 484
  24. The Nickel Arsenide and Related Structures...................... 484
    a. The "Filled NiAs" Structure................................. 487
    b. The WC and AlB_2 Structures............................... 487
  25. The Sodium Chloride and Related Structures...................... 487
    a. The Thallium Iodide Structure............................... 490
    b. The Stannous Sulfide Structure.............................. 491
  26. The $NaHF_2$ and $CsICl_2$ Structures................................ 491
X. Structures Based on 8-Connected Nets.............................. 494
  27. The Body-Centered Cubic and Related Structures.................. 494
  28. The Caesium Chloride Structure................................. 496
XI. Relations between Crystal Structures............................... 499
    a. Structures Closely Related to Simple Structures A_mX_n........... 499
    b. Complex Structures Built of Slices of Simpler Structures.......... 499
    c. Families of Structures with a Common Structural Theme.......... 500

WHI

WHITE, Robert M.　　SUPPLEMENT 15, 1979

Long Range Order in Solids

ROBERT M. WHITE

Xerox Corporation, Palo Alto Research Center, Palo Alto, California

THEODORE H. GEBALLE

*Department of Applied Physics, Stanford University, Stanford, California
and Bell Laboratories, Murray Hill, New Jersey*

Contents

PREFACE ix

I. Phase Transitions and Order Parameters

A. Disorder to Order 1
B. Critical Exponents 14
C. Order Parameters 15
D. Broken Symmetry 36
E. Long Range Order and Dimensionality 42

II. Phenomenological Theories

A. Landau Theory 47
B. Ginzburg–Landau Theory 54
C. Beyond Ginzburg–Landau 78
D. Hydrodynamics 84
E. Time-Dependent Ginzburg–Landau Theory 89

III. The Superconducting Transition Temperature

A. Electron–Phonon Mechanisms 92
B. Excitonic Mechanisms 121

IV. The Curie and Néel Temperatures

A. Direct Exchange 124
B. Superexchange 133
C. Virtual Phonon Exchange 143
D. Coulomb Interactions among Itinerant Electrons 144
E. Indirect Exchange 159
F. Anisotropy 161

V. Experimental Considerations

A. Calorimetry 167
B. Magnetic Susceptibility 170
C. Resistivity Measurements 179
D. Scattering 185
E. Resonance 191
F. Tunneling 197

VI. Systematics of Long Range Order

A. Superconductivity 220
B. Magnetism 247

WHITE, Robert M.

VII.   Impurities and Long Range Order

A.   The Virtual Bound State 266
B.   Manifestations of Local Moments 278
C.   Coexistence of Superconductivity and Ferromagnetism 296
D.   Coexistence of Superconductivity and Charge Density Waves 300

VIII.   Domain Structures

A.   Magnetic Domains 302
B.   Magnetic Domain Walls 307
C.   Magnetic Bubbles 315
D.   The Magneto-Optical Probe 317
E.   Inhomogeneous States in Superconductors 323
F.   Surfaces 337
G.   Inhomogeneities in One and Two Dimensions 345

IX.   Order Amidst Disorder

A.   Introduction 350
B.   Preparation 352
C.   Characterization 354
D.   Superconductivity in Amorphous Metals 364
E.   Amorphous Magnetism 369
F.   Granular Metals 383

Epilogue 391

Author Index 393
Subject Index 401
Substance Index 410

WIGNER, Eugene P.   VOLUME 1, 1955

Qualitative Analysis of the Cohesion
in Metals

EUGENE P. WIGNER AND FREDERICK SEITZ

Princeton University and the University of Illinois

 I. Introduction... 97
 II. Preliminary Observations.. 99
    1. Allotropic Modifications.. 100
    2. Periodic Functions; the Boundary Correction....................... 100
    3. Correlation Effects... 103
    4. Spin-Orbit Interaction.. 106
 III. The Structure and Width of Separated p and d Bands............. 107
 IV. Overlapping Bands... 116
 V. Interpretation of the Cohesive Energies of Metals.................... 119
 VI. Summary... 123
    Appendix: Calculation of the Hole Energies........................... 123

WIL–WOK

WILLIAMS, F.E.: *see* PIPER, W.W.

WOKAUN, Alexander VOLUME 38, 1984

Surface-Enhanced Electromagnetic Processes

ALEXANDER WOKAUN

ETH Zentrum, Physical Chemistry Laboratory
Zurich, Switzerland

I.	Introduction and Survey	224
II.	Surface-Enhanced Raman Scattering: Experimental Observations	226
	1. The Importance of Surface Roughness	227
	2. Experiments on Metal-Island Films and in Ultrahigh Vacuum	227
	3. Inelastic Background Signals in SERS	228
	4. The Wavelength Dependence of SERS	229
III.	Surface-Enhanced Raman Scattering: Theoretical Models	229
	5. Electromagnetic Models	230
	6. Electron–Hole Pair Models	231
	7. The Adatom Model	232
	8. Surface Complexes. Resonant Raman Scattering Induced by Adsorption	233
	9. Modulated Reflectance	234
	10. Image Dipole Mechanism	235
	11. The Electromagnetic Particle Plasmon Model: Mathematical Description	235
IV.	Evidence for the Electromagnetic Model	239
	12. The Distance Dependence of SERS	239
	13. Correlation with Optical Properties	240
	14. The Frequency Dependence of SERS	241
V.	Extensions of the Electromagnetic Model	245
	15. The Lightning-Rod Effect	245
	16. Radiation Damping	248
	17. The Size Effect	254
VI.	Particle Dipolar Interactions	255
	18. The Local Field	256
	19. Shift and Broadening of the Plasmon Resonance	257
	20. Retarded Dipolar Interactions in Regular Particle Lattices	258
	21. Directional Reradiation	261
	22. Simultaneous Resonating of Excitation and Raman Frequencies	263
	23. Determination of Particle Dipole Moments from Reflectivity Measurements	265

WOKAUN, Alexander

VII. Enhanced-Surface Second-Harmonic Generation 269
    24. Enhancement of SHG by Particle Plasmon Resonances 270
    25. Experiments on Ag Island Films 271
    26. Determination of Local Field Factors from Measurements
        of Transmission and Reflection 273
    27. SHG from Regular Ag Particle Arrays 274
VIII. Enhanced Absorption by Adsorbed Dyes 275
    28. Enhanced One-Photon Absorption 276
    29. Enhanced Two-Photon Absorption 279
IX. Contributions to the Enhancement by Other Mechanisms 283
    30. The Distance Dependence of SERS and the Role of Adatoms 284
    31. The Role of Extended Surface Plasmons 285
    32. SERS from Unroughened Metal Surfaces 286
    33. Effects of Chemical Bonding 286
X. Applications of Surface-Enhanced Phenomena; Summary 287
    34. Surface Analytics Using Enhanced Raman Scattering 288
    35. Surface Energy Conversion 290
    36. Summary and Outlook 293

WOLF, E.L.   VOLUME 30, 1975

Nonsuperconducting Electron Tunneling Spectroscopy

E.L. WOLF

*Research Laboratories, Eastman Kodak Company, Rochester, New York and
Cavendish Laboratory, University of Cambridge, Cambridge, England*

  I. Introduction 2
 II. Basic Concepts and Methods 4
    1. Introduction: Exponentially Decaying Waves 4
    2. Noninteracting Final States in the One-, Two-, and Three-Dimensional
        Cases 7
    3. Forms of Tunneling Transport and Spectroscopy 11
    4. Fermi Wavelength versus Coherence Length 20
    5. Recent Experimental Methods, Including Spin-Polarized Tunneling . . . 22
III. Crystalline Barriers: Tunnel Currents, One-Electron Theory, and the E–κ
    Relation 28
    6. Single Crystal GaSe Barriers: Quantitative Metal–Insulator–Metal
        Tunnel Currents 28
    7. Single Crystal Schottky Barriers 31
IV. Final State Spectroscopy, Including Landau Levels and Electron Standing
    Waves 33
    8. Surface and Bulk Landau Levels in Metal–Semiconductor Tunneling . . 33
    9. Electron Standing-Wave Splittings in Metal Films 37
    10. Other Band Structure Effects and Noneffects 41
    11. Spin Polarization in Ferromagnets 45

WOL

WOLF, E.L.

12.	Electron–Phonon Self-Energy Effects	48
13.	Anomalous Tunneling Near the Metal–Semiconductor Transition	51
14.	Tunneling Into Amorphous Semiconductors	54
15.	Proximity Effect Studies	56
V.	Inelastic Assisted Tunneling: Spectroscopy of Thresholds	60
16.	Phonons and Impurity Vibrations	60
17.	Electronic Excitations: Plasmons	63
18.	Magnetic Excitations: Magnons	66
19.	Real-Intermediate-State and Resonant-Inelastic Tunneling	67
VI.	Elastic Assisted Tunneling, Principally Kondo Scattering	69
20.	The Kondo Elastic Scattering Peak	69
21.	$S = \frac{1}{2}$ Hydrogenic Moments in Schottky Barriers	72
22.	Transition Metal Moments in Metal–Insulator–Metal Junctions	76
23.	Discussion	78
VII.	Unusual Materials and Effects	80
24.	Semiconductors and Transition Metal Oxides	80
25.	Metals and Semimetals	81
26.	Miscellaneous Effects	83
VIII.	Tunneling Theory	85
27.	Several Working Calculations	85
28.	New Conceptual Bases for Tunneling	87
IX.	Conclusion	89

WOLF, H.C. VOLUME 9, 1959

The Electronic Spectra of Aromatic Molecular Crystals

H.C. WOLF

II. Physikalisches Institut der Technischen Hochschule, Stuttgart, Germany

I.	Introduction	1
II.	Survey of the Spectra of Aromatic Molecules	2
III.	The Experimental Investigation of the Spectra of Crystals of Aromatic Molecules	9
1.	Survey of Older Work	9
2.	Goal of the Newer Work	10
3.	Methods of Procedure	11
4.	Results of Measurement	25
IV.	Theory of the Spectra of Molecular Crystals and Comparison with Experiment	67
5.	Comparison with an "Oriented Gas"	69
6.	Quantitative Treatment of Displacement and Splitting	69
7.	Qualitative Treatment of Additional Crystal Properties	77
V.	Survey	81

WOLFF, P.A.: *see* PLATZMAN, P.M.

WOLLAN, E.O.: *see* SHULL, C.G.

WOODBURY, H.H.: *see* LUDWIG, G.W.

WOODRUFF, Truman O.   VOLUME 4, 1957

The Orthogonalized Plane-Wave Method

TRUMAN O. WOODRUFF

General Electric Research Laboratory, Schenectady, New York

I. Description of the Method	367
II. Survey of Some Applications of the Method	378
III. Example: The Method Applied to Silicon Crystal	384
1. Motivation	384
2. Construction of the Crystal Potential	385
3. Core States	388
4. Orthogonality Coefficients and Fourier Coefficients of Crystal Potential Energy	394
5. Results and Conclusions	400
Appendix I. A Useful Lattice Sum	404
Appendix II. Matrix Elements of the Radial Atomic Hamiltonian Operator for Silicon	406

WOO

WOOTEN, F.   VOLUME 40, 1987

Modeling Tetrahedrally Bonded Random Networks by Computer

F. WOOTEN

Department of Applied Science,
University of California
Davis/Livermore, California

D. WEAIRE

Physics Department,
Trinity College,
Dublin, Ireland

I. A Short History of Random-Network Models   2
   1. Introduction   2
   2. Background: Some Key Questions   4
   3. Model Building for Amorphous Silicon and Germanium   7
   4. Strategy and Tactics for Computer Modeling   10
II. Computer Generation of Continuous Random-Network Models with Periodic
   Boundary Conditions   11
   5. Introduction   11
   6. Guttman's Method   12
   7. Bond Transpositions   15
   8. Definition of Sillium   17
III. Results   24
   9. Initial Randomization   24
   10. Annealing the Randomized Model   29
   11. Characteristics of the Model   30
   12. Viewing the Model: Computer Graphics   33
   13. Relative Energies   34
   14. The Spectrum of Bond-Switch Energies   35
IV. Conclusion   37
   Appendix A. The Radial Distribution Function   39

WU, David T.   VOLUME 50, 1996

Nucleation Theory

DAVID T. WU

Center for Materials Science, Los Alamos National Laboratory, Los Alamos, New Mexico

I. Introduction   38
II. Classical Theory   39
    1. Homogeneous Nucleation.   39
    2. Heterogeneous Nucleation   51
III. Kinetic Approach   57
    3. Kinetic Potential   59
    4. Kinetic Modification of Gibbs Potential   62
IV. Rate Equation   93
    5. Time-Lag   95
    6. Transient Solution with Method of Moments 101
V. Fokker-Planck Equation 110
    7. Continuum Approximation 110
    8. Time-Lag and Moments 126
VI. Topics in Multicomponent Nucleation 135
    9. Preliminary 135
    10. Kinetic Considerations 139
    11. Dimensional Reduction 144
Appendix A 155
Appendix B 156
Appendix C 160
Appendix D 164
Appendix E 165
Appendix F 168
Appendix G 170
Appendix H 172
Appendix I 173
Appendix J 173
Appendix K 175
Appendix L 178
Appendix M 181
Appendix N 182
Appendix O 182
Appendix P 183
Appendix Q 184
Appendix R 186

g FACTORS AND SPIN-LATTICE RELAXATION
OF CONDUCTION ELECTRONS

Y. YAFET

Bell Telephone Laboratories, Incorporated, Murray Hill, New Jersey

I. Introduction...	2
1. Scope of Article	3
II. Spin-Orbit Coupling...	3
2. Spin-Orbit Coupling in Isolated Atoms	3
3. Kramers Degeneracy and Its Effect on Matrix Elements...............	11
III. The *g* Factor: Theory...	15
4. The *g* Factor at a Nondegenerate Band Edge	15
5. The Magnetic Moment at a Point of No Degeneracy...................	22
6. Wave Functions and Selection Rules for an Approximate Hamiltonian....	24
7. The *g* Factors at a Point of Degeneracy: The Valence Band Edge of Germanium	29
8. Position of the Resonance Line......................................	35
IV. The *g* Factor: Experimental Results and Comparison with Theory..........	39
9. Indium Antimonide..	39
10. Bismuth..	45
11. Silicon...	49
12. Germanium...	54
13. Graphite...	56
14. The Alkali Metals...	61
15. Other Metals...	65
16. The Sign of the *g* Shift..	66
V. Spin-Lattice Relaxation in Nondegenerate Bands.........................	67
17. The Relaxation Times T_1 and T_2..............................	67
18. The Theory of Spin-Lattice Relaxation in Metals.....................	70
19. The Matrix Element for Spin Reversal...............................	74
20. The Relaxation Time in Semiconductors.............................	80
21. The Relaxation Time in Sodium.....................................	81
22. Spin-Lattice Relaxation in High Magnetic Fields	84
23. Relaxation by Impurity Scattering.................................	87
24. Relaxation in Shallow Donor States................................	88

YEOMANS, Julia   VOLUME 41, 1988

The Theory and Application of Axial Ising Models

JULIA YEOMANS

Department of Theoretical Physics
Oxford, England

I. Theory 151
  1. Introduction 151
  2. The ANNNI Model 153
  3. Theoretical Techniques 159
  4. Related Models 168
  5. The Chiral Clock Model 176
II. Experimental Applications 180
  6. Binary Alloys 180
  7. Polytypism 188
  8. Magnetic Systems 196
  9. Conclusion 199

YONEZAWA, Fumiko   VOLUME 45, 1991

Glass Transition and Relaxation of Disordererd Structures

FUMIKO YONEZAWA

Department of Physics
Keio University, Yokohama, Japan

I. Introduction 179
II. General Discussions 181
  1. What Is a Glass Anyway? 181
  2. Models of Amorphous Structures 185
III. Glass Transition 186
  3. Experimental Results 186
  4. Theoretical Approaches 192

YON–YU

YONEZAWA, Fumiko

IV. Computer Simulations of Glass Transition 197
    5. Merits and Demerits 197
    6. Molecular Dynamics Methods 197
    7. Simulations under Constant Pressure 201
V. Simulations of Glass Transition at Constant Pressure 202
    8. Quenching of a Lennard-Jones System 202
    9. Physical Properties through Glass Transition 205
    10. Glass Transition versus Crystallization 219
    11. Microscopic Structure Parameters 224
VI. Fluctuation and Relaxation in Disordered Systems 234
    12. Relaxation of a Supercooled Liquid and a Glass 234
    13. Relaxation of a Glass Prepared with Lower Quench Rate 244
    14. Visual Presentation of Atomic Configurations 249
VII. Summary and Future Work 254

YU, Edward T. VOLUME 46, 1992

Band Offsets in Semiconductor Heterojunctions

EDWARD T. YU

JAMES O. MCCALDIN

THOMAS C. MCGILL

California Institute of Technology
Pasadena, California

I. Introduction 2
II. Theories and Empirical Rules 8
    1. Empirical Rules 8
    2. Band Offsets Calculated as Bulk Parameters 12
    3. Self-Consistent Calculations for Specific Interfaces 21
    4. Comparisons among Theories 24
III. Experimental Techniques 26
    5. Optical Spectroscopy 27
    6. XPS and Related Techniques 29
    7. Electrical Measurements 31
IV. III–V Material Systems 33
    8. GaAs/AlAs 33
    9. InGaAs/InAlAs/InP 41
    10. InAs/GaSb/AlSb 52
V. II–VI Material Systems 59
    11. HgTe/CdTe 60
    12. CdSe/ZnTe 66
    13. Other II–VI Heterojunctions 69

YU, Edward T.

VI.		Strain Effects in Lattice-Mismatched Heterojunctions	74
	14.	Influence of Strain on Electronic Structure	74
	15.	Critical Thickness for Strain Relaxation	78
	16.	Si/Ge	81
	17.	InGaAs/GaAs	92
	18.	Other Lattice-Mismatched Heterojunctions	98
VII.		Heterovalent Material Systems	103
	19.	GaAs/Ge	103
	20.	GaAs/ZnSe	109
	21.	AlSb/GaSb/ZnTe	114
	22.	Other Heterovalent Interfaces	119
VIII.		Discussion	130
	23.	Theory versus Experiment in Lattice-Matched Heterojunctions	131
	24.	Strain	143
	25.	Unresolved Issues	144
IX.		Conclusions	145
		Acknowledgments	146

ZAK, J. VOLUME 27, 1972

The *kq*-Representation in the Dynamics of Electrons in Solids

J. ZAK

Department of Physics, Technion—Israel Institute of Technology, Haifa, Israel

I. Introduction	1
II. The *kq*-Representation	4
III. Bloch Electron in the *kq*-Representation	10
IV. The Impurity Problem in Semiconductors	13
V. Bloch Electrons in a Constant Electric Field	17
VI. Bloch Electrons in a Magnetic Field	26
VII. Fundamental Dynamics in the Bloch Theory of Solids	43
VIII. Symmetry of a Bloch Electron in the Presence of an External Magnetic Field	47
IX. Conclusions	52
Appendix A: Wavefunctions in Different Representations	54
Appendix B: Commutator Expansions	55
Appendix C: Transformation Matrix $(\mathbf{kq}|n\mathbf{k}')$ in the *r*-Representation	57
Appendix D: General Formula for the Expansion of the Hamiltonian	58

ZHA–ZHE

ZHANG, Zhe: *see* LIEBER, Charles M.

ZHELUDEV, I.S.   VOLUME 26, 1971

Ferroelectricity and Symmetry

I.S. ZHELUDEV

Institute of Crystallography, Academy of Sciences, Moscow, USSR

Introduction... 429
I. Symmetry.. 430
II. Phase Transitions in Ferroelectrics................................... 435
III. Domains in Ferroelectrics... 448
IV. Antiferroelectrics.. 454
V. Conclusions... 464

ZHELUDEV, I.S.   VOLUME 29, 1974

Piezoelectricity in Textured Media

I.S. ZHELUDEV

Institute of Crystallography, Academy of Sciences, Moscow, USSR

I. Introduction.. 315
II. General Considerations on the Piezoelectric Effect in Textures........... 316
III. Piezoelectric Textures of Rochelle Salt.............................. 323
  1. Texture Preparation... 323
  2. Some Physical Properties.. 325

ZHELUDEV, I.S.

IV. Piezoelectric Textures Containing Barium Titanate...................... 327
   3. Relation between the Properties of a Texture and of the Constituent
     Crystals.. 328
   4. The Properties of Heterogeneous Textures Containing Piezoelectric and
     Nonpiezoelectric Components.................................... 331
   5. Parallel Combination... 336
   6. Series Combination... 337
V. Wood as a Piezoelectric Texture.................................... 341
   7. Quantitative Studies of the Piezoelectric Effect in Wood.............. 342
   8. The Piezoelectric Properties of Wood and Cellulose Substances........ 344
   9. The Supermolecular Structure of Cellulose in the Light of Symmetry and
     Piezoelectric Properties.. 348
VI. The Piezoelectric Properties of Rocks............................... 351
   10. Possible Types of Texture from Quartz and Quartziferous Rocks...... 352
   11. Experimental Data on the Piezoelectric Properties of Rocks........... 354
VII. Some Remarks on the Piezoelectric Effect in Human and Animal Bones.... 358

ZHU, Jian-Gang: *see* BERTRAM, H. Neal

ZIMAN, J.M.: *see* SHAM, L.J.

ZIMAN, J.M. VOLUME 26, 1971

The Calculation of Bloch Functions

J.M. ZIMAN

H.H. Wills Physics Laboratory, University of Bristol, Bristol, England

I. The Basic Problem of Solid State Theory............................... 1
   1. Technique, Art, or Science?.. 1
   2. The Lattice Potential... 4
   3. Relativistic Hamiltonians... 13
II. Mathematical Insights and Physical Concepts.......................... 15
   4. The LCAO Representation... 15
   5. The Cellular Method... 22
   6. The NFE Representation.. 27
   7. d Bands and Resonances.. 33
III. Pseudopotentials... 39
   8. Orthogonalized Plane Waves.. 39
   9. Analytic Pseudopotentials... 47
   10. Model Pseudopotentials... 59
IV. Augmented Plane Waves and Greenians................................. 69
   11. Augmented Plane Waves.. 69
   12. The Greenian (KKR) Method.. 75
   13. The Greenian in Reciprocal Lattice Representation................. 81
   14. The KKR Method in APW Language................................... 84
   15. The KKR Method and the Model Hamiltonian......................... 92
V. Conclusion... 99

ZOK, F.W.: *see* EVANS, A.G.

ZUNGER, Alex　　　　VOLUME 39, 1986

Electronic Structure of 3d Transition-Atom Impurities in Semiconductors

ALEX ZUNGER

Solar Engergy Research Institute
Golden, Colorado

I.	Introduction	276
II.	Phenomenology and Terminology	277
	1. Impurity Oxidation and Charge States	277
	2. Elements of Electronic Structure in T_d Symmetry	282
III.	Observable Phenomena and Data	290
	3. Types of Experiments	290
	4. Stationary, Chemically Specific Techniques	291
	5. Charge-Specific Techniques Involving a Single Electronic State	292
	6. Techniques That Measure Interelectronic Transitions	303
	7. Observations of the Lattice Symmetry around the Impurity	327
	8. Trends in the Data	332
	9. Puzzles and Dual Characteristics of 3d Impurities in Semiconductors	344
IV.	Theoretical Content of Excitation and Ionization Energies	346
	10. Mean-Field, Many-Electron, and Relaxation Contributions	346
	11. Contemporary Approaches to Separation of MF and MC Effects	354
	12. The Effective-Crystal-Field Approach	358
	13. Chemical Trends in the Mean-Field Parameters	361
	14. Chemical Trends in Excitation and Ionization Energies	370
V.	Electronic-Structure Calculations in the Mean-Field Approach	377
	15. The Point-Ion Crystal-Field Model	378
	16. Ligand-Field and Cluster Models	380
	17. Green's-Function Methods	388
VI.	Results of First-Principles Green's-Function Studies	396
	18. One-Electron Impurity Levels and Wave Functions	396
	19. Charge Densities	413
	20. The Self-Regulating Response	417
	21. Effective Electronic Configurations	423
	22. The Bulk Analog of 3d Impurities in Semiconductors	426
	23. Breathing-Mode Lattice Distortions around the Impurity	429
	24. Effects of Spin Polarization	431
	25. g values	441
	26. Donor, Acceptor, and Mott–Hubbard Energies	442
	27. The Exchange-Correlation-Induced Negative "Effective U"	449

ZUN–ZWI

ZUNGER, Alex

ZWICKNAGL, Gertrud: *see* FULDE, Peter

90090

9 780126 077520

ISBN 0-12-607752-5